Lecture Notes in Mathematics

Edited by A. Dold and B. Eckmann

1158

Stochastic Processes – Mathematics and Physics

Proceedings of the 1st BiBoS-Symposium held in Bielefeld, West Germany, September 10–15, 1984

Edited by S. Albeverio, Ph. Blanchard and L. Streit

Springer-Verlag
Berlin Heidelberg New York Tokyo

Editors

Sergio A. Albeverio
Mathematisches Institut, Ruhr-Universität Bochum
4630 Bochum, Federal Republic of Germany

Philippe Blanchard
Ludwig Streit
Fakultät für Physik, Universität Bielefeld
4800 Bielefeld, Federal Republic of Germany

Mathematics Subject Classification (1980): 03xx, 22xx, 28xx, 31xx, 34Bxx, 35xx, 35Jxx, 46xx, 58xx, 60Gxx, 60Hxx, 60Jxx, 60J45, 73xx, 76xx, 81Fxx, 81Gxx, 82xx, 85xx, 93xx

ISBN 3-540-15998-3 Springer-Verlag Berlin Heidelberg New York Tokyo
ISBN 0-387-15998-3 Springer-Verlag New York Heidelberg Berlin Tokyo

Library of Congress Cataloging-in-Publication Data. BiBoS-Symposium (1st: 1984: Bielefeld, Germany) Stochastic processes, mathematics and physics. (Lecture notes in mathematics; 1158) Bibliography: p. 1. Stochastic processes–Congresses. 2. Mathematics–Congresses. 3. Physics–Congresses. I. Albeverio, Sergio. II. Blanchard, Philippe. III. Streit, Ludwig, 1938-. IV. Bielefeld-Bochum Research Center Stochastics. V. Series: Lecture notes in mathematics (Springer-Verlag); 1158.
QA3.L28 no. 1158 [QA274.A1] 510 s [519.2] 85-26088
ISBN 0-387-15998-3 (U.S.)

Printing and binding: Beltz Offsetdruck, Hemsbach/Bergstr.
2146/3140-543210

Preface

The "1st BiBoS Symposium Stochastic processes: Mathematics and Physics" was held at the Center for Interdisciplinary Research, Bielefeld University, in September 1984. It is the first of a series of symposia organized by the Bielefeld - Bochum Research Center Stochastics (BiBoS), sponsored by the Volkswagen Stiftung.

The aim of the topics chosen was to present different aspects of stochastic methods and techniques in a broad field ranging from pure mathematics to various applications in physics. The success of the meeting was due first of all to the speakers: thanks to their efforts it was possible to take recent developments into account and to speed up publication of the written versions of lectures given at the Symposium.

We are also grateful to the staff of ZiF, in particular to Ms. M. Hoffmann, who expertly handled the organization of the meeting, and to Ms. B. Jahns and M. L. Jegerlehner, who prepared the manuscripts for publication.

S. Albeverio, Ph. Blanchard, L. Streit

Bielefeld and Bochum, December 1985

CONTENTS

Preface

The "1st BiBoS Symposium Stochastic processes: Mathematics and Physics" was held at the Center for Interdisciplinary Research, Bielefeld University, in September 1984. It is the first of a series of symposia organized by the Bielefeld - Bochum Research Center Stochastics (BiBoS), sponsored by the Volkswagen Stiftung.

The aim of the topics chosen was to present different aspects of stochastic methods and techniques in a broad field ranging from pure mathematics to various applications in physics. The success of the meeting was due first of all to the speakers: thanks to their efforts it was possible to take recent developments into account and to speed up publication of the written versions of lectures given at the Symposium.

We are also grateful to the staff of ZiF, in particular to Ms. M. Hoffmann, who expertly handled the organization of the meeting, and to Ms. B. Jahns and M. L. Jegerlehner, who prepared the manuscripts for publication.

S. Albeverio, Ph. Blanchard, L. Streit

Bielefeld and Bochum, December 1985

CONTENTS

List of speakers

L. Arnold
E. Carlen
G.F. Dell'Antonio
E.B. Dynkin
D. Dürr
D. Elworthy
H. Föllmer
L. Gross
F. Guerra
J. Hawkes
Y. Higuchi
R. Høegh-Krohn
S. Kusuoka
J. Lewis
T. Lindstrøm
Y. Le Jan
A. Meyer
H. Nagai
M. Pinsky
W.R. Schneider
R. Seneor
B. Souillard
R. Streater
A. Truman

Stochastic Lie group-valued measures and their relations to stochastic curve integrals, gauge fields and Markov cosurfaces

by

Sergio Albeverio[*,#], Raphael Hoegh-Krohn[**,#], Helge Holden[**,#]

ABSTRACT

We discuss an extension of stochastic analysis to the case where time is multi-dimensional and the state space is a (Lie) group. In particular we study stochastic group-valued measures and generalized semigroups and show how they can be obtained by multiplicative stochastic integration from vector-valued stochastic Lévy-Khinchin fields. We also discuss their connection to group-valued Markov cosurfaces and, in the case of 2-dimensional "time", group-valued curve integrals. We analyze furthermore, in the general multi-dimensional case, the relation with curve integrals, connections and gauge fields and mention the application of group-valued Markov cosurfaces to the construction of relativistic fields.

* Mathematisches Institut, Ruhr-Universität, Bochum and Bielefeld-Bochum Stochastics Research Centre, Volkswagenstiftung

** Matematisk Institutt, Universitetet i Oslo, Oslo

\# Université de Provence and Centre de Physique Théorique, CNRS, Marseille

1. Introduction

In the case of "one dimensional time", Markov processes on manifolds have been studied in different connections, both by analytical semigroup methods e.g. [1]-[3] (see also e.g. [4]-[6] and for many references [7]) and by probabilistic methods, e.g. [8] - [10] (see also e.g. [4] - [7], [11] - [13]). The interplay between analytic, probabilistic and geometric problems and methods has been given great attention in recent years (e.g. [14], [15], [7], [16] - [22] and references therein). Markov processes on manifolds have also been used in connection with non relativistic quantum mechanics, see e.g. [6], [7], [15], [29].

The case where the manifold is that of a Lie group is, on one hand, a particular one, but on the other hand, due to the particular geometric structure of Lie groups, it also has pecularities which makes it very worthwhile studying. In fact such studies have been done, particularly in the case of diffusions, see e.g. [10], [13], [22], [23].

An extension of those methods and results to the case where the time parameter is multidimensional, i.e. of random maps from a manifold (or more generally, any measurable space) into another manifold is of great interest, for many reasons. We mention in particular, besides stochastic analysis itself, e.g. [26] - [28], the study of representations of certain infinite dimensional Lie groups and the construction of non commutative distributions, see e.g. [16], [29], the study of non commutative random fields [15] and quantum fields [15], [30] - [32]. In this lecture we shall in particular present a new approach to the construction of the Markov fields in the case where the target manifold is a Lie group. This approach is based on an extension to the case of more dimensional "time parameter" of methods of stochastic analysis on manifolds. It turns out that the random fields which we construct have interesting invariance properties which make it possible to associate to them relativistic fields and, in the case of a 2-dimensional manifold, random connections and stochastic Euclidean gauge fields. The basic object underlying the construction of such Markov fields is an extension of the concept of Markov semigroup to the case of index set a measurable space $(M,\mathcal{B})$, obtaining so called "generalized Markov semigroup" $(p_A, A \in \mathcal{B})$ satisfying essentially

$A \cap B = \emptyset \Rightarrow p_{A \cup B} = p_A * p_B = p_B * p_A$ and a continuity property).

Thus in section 2 of this paper we study such generalized Markov semigroups. We show essentially that they are in one-to-one correspondence with stochastic group-valued multiplicative measures η on $(M,\mathcal{B})$ (with $\eta(A \cup B) = \eta(A)\cdot\eta(B)$ and $\eta(A)$ independent of $\eta(B)$, whenever $A \cap B = \emptyset$, and a continuity condition). As a tool for the construction of such multiplicative measures we study in section 3 stochastic vector-valued (additive) measures and show that they are classified essentially by a Lévy-Khinchin type formula. We use these stochastic vector-valued measures in Sect. 4 to show that, in the case of Lie groups, they are in 1-1 correspondence with stochastic group-valued multiplicative measures, obtained by solving essentially a stochastic differential equation on the Lie group.

Section 5 is concerned with some deterministic notions like multiplicative curve integrals, with their relations to connection and (classical) gauge fields, as well as group-valued (codimension 1) cosurfaces. The corresponding stochastic objects are studied in Section 6. It is shown that they can be realized starting from a generalized Markov semigroup η, constructed as in Sect. 3. The Markov property of such stochastic G-valued cosurfaces is also discussed, as well as the relation to stochastic gauge fields in the case the basic manifold M is $\mathbb{R}^2$. In the case $M = \mathbb{R}^d$ it is also mentioned that to invariant cosurfaces there correspond Markov semigroups on a certain Hilbert space, as well as relativistic quantum fields, associated with hypersurfaces (instead of points).

In Section 7 we discuss how the Markov cosurfaces of Sect. 6 can be obtained as continuum limits from lattice group-valued random fields.

2. Stochastic group-valued multiplicative measures and generalized Markov semigroups

In this section we shall discuss stochastic multiplicative measures and generalized Markov semigroups with applications, in later sections, to the construction of multiplicative integrals.

Let $(M,\mathcal{B})$ be a measurable space (in the applications $(M,\mathcal{B})$ will mainly be $(\mathbb{R}^d, \mathcal{B}(\mathbb{R}^d))$, $\mathcal{B}(\mathbb{R}^d)$ being the Borel subsets of $\mathbb{R}^d$) and let G be a locally compact group. By a stochastic G-valued multiplicative measure η on $(M,\mathcal{B})$ we understand a mapping $A \in \mathcal{B} \to \eta(A)(\omega) \in G$ into G-valued stochastic variables, where ω is a point in some fixed probability space $(\Omega,\mathcal{A},P)$. We require η to satisfy $\eta(\emptyset) = e$ for all $\omega \in \Omega$ and, when $A \cap B = \emptyset$, $\eta(A)$ independent of $\eta(B)$ and $\eta(A \cup B) \overset{\Delta}{=} \eta(A) \cdot \eta(B)$, where $\overset{\Delta}{=}$ means equality in law. Moreover we require continuity in law $\eta(A_n) \to \eta(A)$ as $A_n \downarrow A$. As an example of η we might think of $(M,\mathcal{B}) = (\mathbb{R}^d, \mathcal{B}(\mathbb{R}^d))$, $G = \mathbb{R}$, η the white noise real generalized Gaussian stochastic process on $\mathbb{R}^d$ (s.t. $\eta(A) = \int \eta(x)\chi_A(x)dx$ with $\eta(x)$ Gaussian with mean zero and covariance $E(\eta(x)\eta(y)) = \delta(x-y)$).

Most useful for our purposes are stochastic multiplicative measures η taking values in unimodular locally compact groups and having the property that if $\eta(A) = e$ does not hold P-a.s., then the distribution of $\eta(A)$ is equivalent with Haar's measure on G. We say in this case that η is strongly ergodic. For strongly ergodic non trivial η (i.e. such that for any $A \in \mathcal{B}$ either $\eta(A) \neq e$ or $\eta(M-A) \neq e$) we have that the distribution p_A of $\eta(A)$ is invariant in the sense that its density $\tilde{p}_A$ with respect to Haar measure satisfies $\tilde{p}_A(h^{-1}kh) = \tilde{p}_A(k)$, for Haar-a.e. $h,k \in G$. For this result see [39].

We shall now introduce the concept of generalized Markov semigroup on a locally compact group G. We call so by definition any map p from a measurable space $(M,\mathcal{B})$ into probability measures on G satisfying $p_{A \cup B} = p_A * p_B = p_B * p_A$, if $A \cap B = \emptyset$, $p_{A_n} \to p_A$ weakly as $A_n \downarrow A$.

By the above, any multiplicative G-valued measure η on $(M,\mathcal{B})$ defines a generalized Markov semigroup on G, p s.t. p_A is the law of $\eta(A)$. If η is strongly ergodic and non trivial then it defines an invariant generalized Markov semigroup p_A on G, in the sense that $\tilde{p}_A(h^{-1}kh) = \tilde{p}_A(k)$ for P-a.a.h, $k \in G$, $\tilde{p}$ being the density of p.

Examples 1. For an example of a generalized Markov semigroup on G, let us consider the case $(M,\mathcal{B}) = (\mathbb{R}^+, \mathcal{B}(\mathbb{R}^+))$. Let $(P_t, t \geq 0)$ be a 1-parameter invariant weakly continuous Markov convolution semigroup on G, set $p_{[t_1,t_2]} \equiv P_{t_2-t_1}$, $0 \leq t_1 \leq t_2$ and extend naturally by continuity the definition of p_A to all $A \in \mathcal{B}(\mathbb{R}^+)$. Then p_A is an invariant generalized Markov semigroup on G.

2. We can use the invariant Markov semigroup P_t of Ex. 1 to construct invariant generalized Markov semigroups p_A on an arbitrary σ-finite positive measure space $(M,\mathcal{B},\sigma)$, provided $P_t \to P_\infty$ as $t \to \infty$ weakly (no such restriction is needed if σ is finite). In fact define $p_A \equiv P_{\sigma(A)}$ for any $A \in \mathcal{B}$, it is then easily verified that $(p_A, A \in \mathcal{B})$ has the properties of an invariant G-valued generalized Markov semigroup. In particular M can be a Riemannian manifold and σ the Riemannian volume measure on it, we shall see below (in Sect. 6) the importance of such a case.

There is a converse of the above relation: multiplicative G-valued measure $\Rightarrow$ generalized Markov semigroup, in fact we have the following. Let G be a locally compact separable unimodular group. If $A \to p_A$, $A \in \mathcal{B}$ is a generalized Markov semigroup on G with parameter space $(M,\mathcal{B})$, then there exists a stochastic G-valued multiplicative measure η on $(M,\mathcal{B})$ s.t. p_A is the distribution of $\eta(A)$. η is unique in the sense that for any finite collection $A_i \in \mathcal{B}$, $i = 1,\ldots,n$, the joint distribution of the $\eta(A_i)$ is unique. If p is invariant then η is <u>invariant</u> (in the sense that for any finite $\mathcal{B}$-measurable partition the joint distribution of the random variables $\eta(A_i)$ is the same as that of the random variables $g_i^{-1}\, \eta(A_i) g_i$, for any arbitrary $g_i \in G$). For the proof of these results, based on an extension of Kolmogorov's construction, see [39].

In particular we get that any strongly ergodic non trivial multiplicative measure is invariant.

In the situation of the above example 2 we have easily that, due to the invariance of P_t, p_A is invariant und hence the multiplicative measure η constructed from it (unique, up to equivalence, as proven in [39], Theor. 2.15, and such that the distribution of $\eta(A)$ is $P_{\sigma(A)}$) is also invariant.

We shall see in Sect. 6 how we can associate multiplicative G-valued integrals to multiplicative G-valued measures, and then use the results of this section to construct the connection and gauge fields mentioned in Sect. 1. However, for the discussion of this association we need first to examine so called stochastic vector valued measures, and this we do in the next section.

3. Stochastic vector valued measures

Let V be a real finite dimensional vector space and let $(M,\mathcal{B})$ be a measurable space. A stochastic V-valued measure ξ on $(M,\mathcal{B})$ is by definition a stochastic V-valued multiplicative measure on $(M,\mathcal{B})$ in the sense of Sect. 2, with G equal to the abelian group V. ξ has then the properties that if $A \cap B = \emptyset$ then $\xi(A)$ and $\xi(B)$ are independent and $\xi(A \cup B) = \xi(A) + \xi(B)$ and $\xi(A_n) \to \xi(A)$ in law if $A_n \downarrow A$. As an example we might take $(M,\mathcal{B}) = (\mathbb{R}^d, \mathcal{B}(\mathbb{R}^d))$, $V = \mathbb{R}^s$, ξ the $\mathbb{R}^s$-valued white noise generalized process. Another example is given by $(M,\mathcal{B}) = (\mathbb{R}^+, \mathcal{B}(\mathbb{R}^+))$, V = g, the Lie algebra of some Lie group G, ξ the white noise generalized process in g.

Let us remark that our stochastic vector valued measures are related, but not identical, with those encountered in the literature on processes with independent increments, see e.g. [40] - [46].

It is easily shown [39] that for any stochastic V-valued measure ξ we have

$$E(e^{i<p,\xi(A)(\cdot)>}) = e^{\int_A \mu(p;dm)}$$

for any $p \in V'$, $A \in \mathcal{B}$, with $\mu(p;dm)$ for any $p \in V'$ a measure on $(M,\mathcal{B})$. Moreover ξ has a decomposition $\xi = \xi_c + \xi_d$ such that $\xi_d(A) \equiv \xi(A \cap D) = \sum_{x_i \in A \cap D} \xi(\{x_i\})$ $\xi_c(A) \equiv \xi(A \cap (M-D))$, with D an at most countable subset of M (at least when (M,) is topological). The V-valued random variables $\xi_i \equiv \xi(\{x_i\})$ are independent. One has

$$E(e^{i<p,\xi_c(A)(\cdot)>}) = e^{\int_A f(p;m)\mu(dm)} , \qquad (3.1)$$

with f(p,m) of Lévy-Khinchin's type, i.e.

$$f(p;m) = \int_{0<\|\alpha\|\leq 1} (e^{i<p,\alpha>} - 1 - i<p,\alpha>)\, \nu(d\alpha,m) +$$

$$+ \int_{\|\alpha\|>1} e^{i<p,\alpha>} \nu(d\alpha,m) - \frac{1}{2} <p,A(m)p>,$$

$<,>$ resp. $\| \ \|$ being the scalar product resp. norm in V, A(m) is, for μ-a.e. m, a(dim V)x(dim V) positive definite symmetric matrix, ν a σ-finite positive measure on V - {0} s.t. $\int_{0<\|\alpha\|\leq 1} \|\alpha\|^2 \nu(d\alpha,m)$, $\int_{\|\alpha\|>1} \nu(d\alpha,m)$ and $<p,A(m)p>$ are all in $L^1(\mu)$.

Viceversa, given the r.h.s. of (3.1), one can define a stochastic V-valued measure $\xi_c(\cdot)$ s.t. (3.1) holds.

For these results, somewhat related to [42], [46], [47], see[39].

By the essential equivalence of stochastic multiplicative measures and generalized semigroups on G, discussed in Sect. 2, we can express the above results for multiplicative measures on V = G in terms of results on generalized Markov semigroups on V = G, i.e. for any such generalized semigroup (p_A, $A \in \mathcal{B}$) there exists an at most countable subset D of M s.t. $p_{\{x\}} = \delta_0$ for any $x \in M-D$ and $p_A = p_A^c * p_A^d$, with $p_A^c \equiv p_{A\cap(M-D)}$,

$$p_A^d \equiv p_{A\cap D} = \mathop{*}_{x_i \in A\cap D} p_{\{x_i\}} ,$$
$$\int e^{i<p,q>} p_A^c(dq) = e^{\int_A f(p,m)\mu(dm)} \tag{3.2}$$

with f, μ as above. And viceversa, the r.h.s. of (3.2) defines a generalized Markov semigroup p_A^c with $p_{\{x\}}^c = \delta_0$.

Our aim in the next section is to take V to be the Lie algebra g of G and then put in relations the stochastic g-valued measures of this section (and the associated generalized Markov semigroups on g) with the G-valued multiplicative integrals of Sect. 2 (and associated generalized Markov semigroups). The essential idea in this is to use (multiplicative) stochastic equations (passing e.g. from the stochastic g-valued measure "white noise in g" to the corresponding G-valued multiplicative integrals defined in terms of Brownian motion on G).

4. Stochastic Lie-group-valued multiplicative measures and integrals

Let G be a connected Lie group with Lie algebra g and let $(M,\mathcal{B})$ be a measurable space. Let η be a stochastic G-valued multiplicative measure on $(M,\mathcal{B})$ and let φ be a positive measurable function on M. It is easily seen, see [39] for details, that

$$t \to \eta_{s,t}(\varphi) \equiv \eta(\varphi^{-1}[s,t]) \tag{4.1}$$

is, for any $s \leq t$, left continuous in law and has the "independent random measures" property in the sense that, whenever $s_1 \leq t_1 \leq s_2 \leq t_2$, $\eta_{s_1,t_1}(\varphi)$ and $\eta_{s_2,t_2}(\varphi)$ are independent and $\eta_{s_1,s_2}(\varphi) \overset{\Delta}{=} \eta_{s_1,t_1}(\varphi)\, \eta_{t_1,s_2}(\varphi)$.

Let us for a moment consider the case where $t \to \eta_{0,t}(\varphi)$ is the left Brownian motion process in G; this can be realized by taking e.g. $M = \mathbb{R}^+$,

$\mathcal{B} = \mathcal{B}(\mathbb{R}^+)$, $\varphi(s) = s, \forall s \in \mathbb{R}^+$, $\eta([0,t]) = b(t)$, with $b(t)$, $t \in \mathbb{R}^+$ a left Brownian motion on G, since then $\eta_{0,t}(\varphi) = b(t)$. In this case the multiplicative stochastic differential equation

$$\eta_{0,t}(\varphi)^{-1}\, d\eta_{0,t}(\varphi) = \xi(t), \tag{4.2}$$

with $\xi(t)$ the white noise generalized process in g, with initial condition $\eta_{0,0}(\varphi) = e \in G$, has a unique non anticipating solution, as can be proven by well known discrete time approximation methods, see e.g. [10], [13].

We can rewrite (4.2) in the integral form, for any $I \in \mathcal{B}(\mathbb{R}^+)$:

$$\int_I \eta_{0,t}(\varphi)^{-1}\, d\eta_{0,t}(\varphi) \equiv \int_I \xi(t) \equiv \xi(I), \tag{4.3}$$

where $\xi(I)$ is the evaluation at I of the white noise measure on g.

The reason the methods for solving (4.2) resp. (4.3) work is essentially that the Brownian motion process has independent increments. By using similar methods, exploiting the above properties of $\eta_{s,t}(\varphi)$, it is possible to give a meaning to the left hand side of (4.3) in the general case where $\eta_{0,t}(\varphi)$ is defined by (4.1) through $\eta(\varphi^{-1}[s,t])$.

Define then

$$\xi(I,\varphi) \equiv \int_I \eta_{0,t}(\varphi)^{-1}\, d\eta_{0,t}(\varphi). \tag{4.4}$$

As in the above case of Brownian motion on G, one can easily show that $I \to \xi(I,\varphi)$ is a stochastic g-valued measure on $(\mathbb{R}^+, \mathcal{B}(\mathbb{R}^+))$.

Moreover one has the transformation law, for any ψ positive measurable on $\mathbb{R}^+$,

$$\xi(I,\psi(\varphi)) = \xi(\psi^{-1}(I),\varphi),$$

as easily proven by a change of variables.

If $(M,\mathcal{B})$ is a standard Borel space we can choose φ to be such that $\mathcal{B} = \varphi^{-1}\mathcal{B}(\mathbb{R}^+)$ and for $A = \varphi^{-1}(I)$, $A \in \mathcal{B}$, $I \in \mathcal{B}(\mathbb{R}^+)$ we can define $\tilde{\xi}(A) \equiv \xi(I,\varphi)$.

By using the above transformation property one sees easily that $\tilde{\xi}(A)$ depends only on A, but not on I und φ.

$\tilde{\xi}$ is a g-valued stochastic measure and if η is strongly ergodic (and non trivial) then $\tilde{\xi}$ is invariant (in the sense of finite dimensional joint distributions) under the adjoint action of G on its Lie algebra.

Using then the structural results of Sect. 3, we have that $\tilde{\xi}$ has a decomposition $\tilde{\xi} = \tilde{\xi}_c + \tilde{\xi}_d$, with $\tilde{\xi}_d(\cdot) \equiv \xi(\cdot \cap D) = \sum_{x_i \in \cdot \cap D} \xi(\{x_i\})$, $\tilde{\xi}_c(\cdot) \equiv \xi(\cdot \cap (M-D))$,

with D at most contable (if, e.g., $(M,\mathcal{B})$ is topological), ξ_i g-valued independent random variables, and $E(e^{i<p,\tilde{\xi}_c(A)>})$ given by (3.1), with $V = g$.

This then gives a construction and characterization of a stochastic g-valued measure starting from a stochastic multiplicative G-valued measure.

Conversely, let us start from a stochastic g-valued measure ξ on a standard Borel space $(M,\mathcal{B})$, invariant under the adjoint action of G (in the sense of finite dimensional joint distributions).

Then to it one can associate a Markov process η_t^I on G, left and right invariant, and with independent right increments, which is the unique non anticipating solution of the stochastic differential equation

$$(\eta_t^I)^{-1}\, d\eta_t^I = \chi_I(t)\, d\xi(\varphi<t), \qquad (4.5)$$

with initial condition $\eta_o^I = e \in G$, with $I \in \mathcal{B}(\mathbb{R}^+)$ arbitrary.

In [39] it is shown how one can use this stochastic process to define a generalized Markov semigroup $\tilde{p}_A$ or $(M,\mathcal{B})$ which is G-invariant. In fact $\tilde{p}_A$ can be defined, using the standard Borel space structure of $(M,\mathcal{B})$ and a measurable mapping φ such that $A = \varphi^{-1}(I)$, $I \in \mathcal{B}([0,1])$, as the distribution p_I of $\eta(I) \equiv \eta_1^I$.

By the results of Sect. 2 we can associate to $\tilde{p}$ a G-valued multiplicative G-invariant measure $\tilde{\eta}$ on $(M,\mathcal{B})$ whose distribution is precisely $\tilde{p}$.

The relation between ξ and $\tilde{\eta}$ is simply

$$\xi(A) = \int_{\varphi(A)} \tilde{\eta}(\varphi < t)^{-1} \, d\tilde{\eta}(\varphi < t) \tag{4.6}$$

i.e., with $A_t = \varphi^{-1}([0,t))$:

$$\xi(A_t) = \int_0^t \tilde{\eta}(A_s)^{-1} \, d\tilde{\eta}(A_s). \tag{4.7}$$

Infinitesimally this can be written as

$$\xi(d\,A_t) = \tilde{\eta}(A_t)^{-1} \, \tilde{\eta}(d\,A_t)$$

with initial condition $\tilde{\eta}(A_0) = e \in G$.

We have $\tilde{\eta}(A_t) \stackrel{\Delta}{=} \eta_t^I$. $\tilde{\eta}$ depends on ξ but not on φ.

In this way we have shown how, essentially, stochastic invariant g-valued measures ξ determine stochastic G-valued multiplicative invariant measures η and the associated invariant generalized Markov semigroups, and viceversa.

In the Section 6 we shall see how we can use stochastic G-valued multiplicative invariant measures to construct G-valued stochastic multiplicative cosurfaces and curve integrals, the deterministic version of which we study in Sect. 5, together with their relations to connections and gauge fields.

5. G-valued cosurfaces, multiplicative curve integrals, connections, gauge fields

We shall consider smooth curves on a connected manifold M and we shall introduce a composition law for them. A smooth curve c on M is by definition the image in M of a smooth injection $t \mapsto c(t)$ of the interval [0,1], with orientation inherited from the natural one on [0,1]. We call S(M) the family of all smooth curves on M. We call $c(0) \equiv c^-$ and $c(1) \equiv c^+$ the endpoints of c. We denote by c^{-1} the curve in S(M) given by the injection $c^{-1}(t) = c(1-t)$. For any 2 "consecutive" curves $c_i \in S(M)$, i = 1,2 with $c_1^+ = c_2^-$ we define a "product" ("juxtaposition") $c_1 \cdot c_2$ as the subset of M given by the injection $c(t) \equiv c_1(2t)$ for $0 \leq t \leq \frac{1}{2}$, $c(t) \equiv c_2(2t-1)$ for $\frac{1}{2} \leq t \leq 1$. The so defined product is associative. We call a subset c of M a piecewise smooth curve if c is of the form $c = c_1 \ldots c_k$ for some $c_i \in S(M)$. We call PS(M) the set of all piecewise smooth curves. The above product has a natural extension in PS(M). We can look upon points $x \in M$ as elements in S(M) and PS(M), namely given by the constant curve $x(t) = x$, $0 \leq t \leq 1$. PS(M) becomes a partial group with partial unit element any $x \in M$. We call $S_x(M)$ resp. $PS_x(M)$ the subset of S(M) resp. PS(M) consisting of curves c("loops") s.t. $c^- = c^+ = x$.

$PS_x(M)$ is then a group with unit element x.

Let G be a group. A G-valued multiplicative curve integral on M is a mapping m from PS(M) into G which is a partial homomorphism, in the sense that for any $c_i \in PS(M)$ for which $c_1 \cdot c_2$ is defined $m(c_1 \cdot c_2) = m(c_1) \cdot m(c_2)$ (where on the right hand side the product is in G) and $m(c^{-1}) = m(c)^{-1}$, for all $c \in PS(M)$ (where $m(c)^{-1}$ denotes the inverse of m(c) in G), and $m(x) = e$ for any $x \in M$, where e denotes the unit element in G.

Remark We shall see below that for M a 2-dimensional oriented manifold a G-valued multiplicative curve integral coincides with a G-valued cosurface in the sense of [33].

Let us denote by m(M,G) the set of all G-valued multiplicative curve integrals. Let $c \in S_x(M)$ and define the family $\tilde{c}(s)$, $0 \leq s \leq 1$ of 1-dimensional subsets of M, with parameter t, by $\tilde{c}(s)(t) \equiv c(s\,t)$, $0 \leq t \leq 1$. $\tilde{c}(s)$ is thus, for fixed s, a smooth curve ($\tilde{c}(s) \in S(M)$), starting at $c(0) = x$ and ending at $c(s)$.

We have $\tilde{c}(0)(t) = x$ for all t i.e. $\tilde{c}(0)$ is the constant x and $\tilde{c}(1)(t) = c(t)$ i.e. $\tilde{c}(1)$ is the curve $c \in S_x(M)$. $\tilde{c}(s)(\cdot)$ is a curve "describing the growth of $c(\cdot)$ from x to c(s)". We denote by $\tilde{c}$ the family of curves $\tilde{c}(s) \in S(M)$, $0 \leq s \leq 1$. We can naturally extend the definition of the operation $\sim$ to all curves $c \in PS_x(M)$, obtaining $\tilde{c}(s) \in PS(M)$. For $c \in PS_x(M)$ and $m \in m(M,G)$ we have a natural definition of $m(\tilde{c}(s))$, since $\tilde{c}(s) \in PS(M)$, for all $0 \leq s \leq 1$.

Let G be a (connected) Lie group. We call $m \in m(M,G)$ smooth if $s \to m(\tilde{c}(s))$ is piecewise smooth mapping from [0,1] into G. We shall denote by $m^\infty(M,G)$ the set of all smooth $m \in m(m,G)$.

Now let G be in addition simply connected and let $m \in m(M,G)$. Then the tangent vector $dm(\tilde{c}(s))$ of the curve $s \to m(\tilde{c}(s))$ in G, computed at $m(\tilde{c}(s)) \in G$, exists. It is a vector in the tangent space $T_{m(\tilde{c}(s))}G$. Let $m(\tilde{c}(s))^{-1}\, dm(\tilde{c}(s))$ be the vector in the Lie algebra $g \cong T_eG$ of G such that the transport of it by the trivial left invariant connection acting from T_eG to $T_{m(\tilde{c}(s))}G$ is $dm(\tilde{c}(s))$. Since $s \to m(\tilde{c}(s))$ is piecewise smooth, we can define $\int_0^1 m(\tilde{c}(s))^{-1}\, dm(\tilde{c}(s)) \equiv \chi_m(c)$, and this is easily seen to be independent of the injection used to define c. The map $c \to \chi_m(c)$ from PS(M) to g has the properties of a g-valued <u>additive curve integral</u>, in the sense that $\chi_m(c_1 \cdot c_2) = \chi_m(c_1) + \chi_m(c_2)$, whenever $c_1 \cdot c_2$ is defined, $\chi_m(c^{-1}) = -\chi_m(c)$, and $\chi_m(c_o) = 0$, whenever c_o is a constant curve. For details see [48]. Since $m \in m^\infty(M,G)$ we have that $s \to \chi_m(\tilde{c}(s))$ is smooth in g whenever $c \in S(M)$.

Conversely for any smooth g-valued additive curve integral there is a family $m_\chi(\tilde{c}(s))$, $0 \leq s \leq 1$ of smooth G-valued multiplicative curve integrals s.t. $c \to m_\chi(\tilde{c}(1))$ is a smooth G-valued multiplicative curve integral on G; $m_\chi(\tilde{c}(s))$ is the unique solution of the differential equation $m_\chi(\tilde{c}(s))^{-1}\, dm_\chi(\tilde{c}(s)) = d\,\chi(\tilde{c}(s))$, $0 \leq s \leq t$ with initial condition $m_\chi(\tilde{c}(0)) = e$.

For any smooth g-valued 1-form ω on M (given in local coordinates x_μ, $\mu = 1,\dots,d$ on M by $\omega_x = \sum_\mu a_\mu(x)dx_\mu$, with dx_μ a basis of T_x^*M and $a_\mu \in C^\infty(M,g)$) and any $c \in PS(M)$ we have that $\chi_\omega(c) \equiv \langle\omega,c\rangle \equiv \int_c \omega$ is well defined. χ_ω is then a special smooth g-valued additive curve integral and by the above one get from it a smooth G-valued multiplicative integral $m_\chi(\tilde{c}(1)) \equiv m_\omega(c)$, which we call <u>the multiplicative G-valued</u>

integral associated with ω.

In local coordinates the covariant derivative $D(\omega)_\mu f$ of $f \in C^\infty(M,g)$ with respect to ω in the direction dx_μ is given by

$$(D(\omega)_\mu f)(x) \equiv \frac{\partial}{\partial x_\mu} f(x) + (ada_\mu(x))f(x), \tag{5.1}$$

with $adA(B) \equiv [A,B]$, the Lie scalar product in g. The covariant exterior derivative $D(\omega)\,\sigma$ of a smooth g-valued one form σ with local representation $\sigma_x = \sum_\mu b_\mu(x)\, dx_\mu$ is given in local coordinates by

$$D(\omega)\sigma = \sum_{\mu,\nu} (D(\omega)\sigma)_{\mu,\nu}(x)\, dx_\mu \wedge dx_\nu,$$

where $(D(\omega)\sigma)_{\mu\nu}(x) = [\frac{\partial}{\partial x_\mu} + ada_\mu(x)]\, b_\nu(x) - [\frac{\partial}{\partial x_\nu} + ada_\nu(x)]\, b_\mu(x)$.

The curvature $F(\omega)$ corresponding to the smooth 1-form ω is the g-valued 2-form $F(\omega)$ given by

$$F(\omega) \equiv D(\omega)\omega.$$

Thus, in local coordinates

$$(F(\omega))_{\mu,\nu}(x) = \frac{\partial}{\partial x_\mu} a_\nu(x) - \frac{\partial}{\partial x_\nu} a_\mu(x) + [a_\mu(x),\, a_\nu(x)]. \tag{5.2}$$

The curvature given by ω is related to the multiplicative curve integral m_ω associated with ω by the formula $dm_\omega(\partial S^x_t)|_{t=0} = d(\int_{S^x_t} F(\omega))|_{t=0}$ i.e.

$$m_\omega(\partial dS_x) = \langle F(\omega),\, dS_x\rangle, \tag{5.3}$$

where dS is a 2-dimensional surface element of M at x, associated with a family S^x_t, $0 \leq t \leq T$ of 2-dimensional submanifolds of M with smooth boundary s.t. ∂S^x_t is a loop starting and ending at x, $S^x_{t_1} \subset S^x_{t_2}$ for $t_1 < t_2$, $S^x_t \downarrow x$ as $t \downarrow 0$. $\langle F(\omega), dS_x\rangle$ means the evaluation of the 2-form $F(\omega)$ at the surface element dS_x.

(5.3) expresses the fact that if S_x is a smooth 2-dimensional surface with boundary ∂S_x then

$$\lim_{|S_x|\to 0} \exp\left(\int_{S_x} F(\omega)\right) = \lim_{|S_x|\to 0} m(\partial S_x),$$

where $|S_x|$ is the area of S_x.

What ist the relation between multiplicative G-valued curve integrals and connections on $M \times V$, with V a suitable finite dimensional vector space?

If τ is a connection on the trivial vector bundle $E = M \times V$ and $c \in PS_x(M)$, let $m(c)$ be the element of GL(V) s.t. $\tau_{(x,v)}(c)^+ = (c^+, vm(c))$, with $vm(c)$ the image in V of $v \in V$ under $m(c)$, $(c^+, vm(c))$ being an element in E. Then $c \to m(c)$ is a multiplicative GL(V)-valued integral. Viceversa, if m is a GL(V)-valued multiplicative integral then $\tau_{(x,v)}(c)(t) \equiv (c(t), vm(\tilde{c}_t))$, with $\tilde{c}_t(s) \equiv c(ts)$, defines a connection on E [48].

In the physical literature gauge fields are defined as connections on suitable vector bundles (e.g. associated to a faithful finite dimensional representation of a compact group G). When the bundle is trivial (which is e.g. the case when the underlying manifold M is $\mathbb{R}^d$), a gauge field can be identified with a $\dot{g}$-valued one-form on M. In this case we can take $V = g$ and characterize, using the above results, the gauge field by a G-valued multiplicative integral.

In the next sections we shall see how to construct in the case $d = 2$ suitable stochastic multiplicative G-valued curve integrals, which in particular yield (Euclidean, stochastic) gauge fields.

First however let us introduce the concept of G-valued <u>cosurface</u>, discussed in [33], [48], and used in next section in a stochastic sense.

Let $d \geq 2$ and let $PS^d(M)$ be the family of all oriented piecewise smooth connected (d-1)-dimensional hypersurfaces on M, i.e. all piecewise smooth injections $[0,1]^{(d-1)} \to M$. For $d = 2$ we have $PS^2(M) = PS(M)$, as defined above.

For $d = 2$ we have already defined the partial product $c_1 \cdot c_2$ of two elements of $PS_d(M)$. For $d > 2$ we define $c_1 \cdot c_2$ iff $c_1 \cap c_2$ is (d-2)-dimensional and contained in $\partial c_1 \cup \partial c_2$, the orientation of $c_1 \cap c_2$ as part of c_1 opposite to the one of the same set as part of c_2. We then define $c_1 \cdot c_2$ to be $c_1 \cup c_2$ as a set, with orientation the one inherited from the ones of c_1 and c_2.

Let $c_1, \ldots, c_n \in PS^d(M)$ be such that $c_1 \cdot c_2$, $(c_1 \cdot c_2) \cdot c_3, \ldots, ((c_1 c_2) c_3 \ldots c_n)$ is defined, and define recursively $c_1 \ldots c_n = (c_1 \ldots c_{n-1}) \cdot c_n$. Let $\sum_M$ be the set of all such products. For any $c \in \sum_M$ we define c^{-1} to be equal to c, but with opposite orientation on M.

(For $d = 1$, $\sum_M$ can be simply identified with the set of all points of M). We define a G-valued cosurface on M as a map from $\sum_M$ into G, with:

$$m(c_1 \cdot c_2) = m(c_1) \cdot m(c_2)$$

$$m(c) = m(c)^{-1}.$$

For $d = 2$ we have that a G-valued cosurface is a G-valued multiplicative curve integral, whenever $m(x) = e$, for x the constant curve coinciding with $x \in M$ (e the unit in G).

6. Markov cosurfaces from generalized Markov semigroups, stochastic multiplicative G-valued measures, stochastic curve integrals and associated relativistic quantum fields

Let $(M, \mathcal{B})$ be an oriented Riemannian manifold of dimension d. Let G be an unimodular locally compact separable group, as in Sect. 2.

Let $\sum_M$ be as in Sect. 5.

Let $\Gamma_{M,G}$ be the set of all G-valued cosurfaces on M, with its natural topology, making all cosurfaces m(c) measurable.

A stochastic G-valued cosurface m on M is by definition a measurable map m from some probability space $(\Omega, \mathcal{A}, P)$ into $\Gamma_{M,G}$. For $d = 2$ we also call a stochastic G-valued cosurface a stochastic G-valued multiplicative curve integral.

Remark For $d = 1$, $M = \mathbb{R}$ a stochastic cosurface is just a C-valued stochastic process indexed by $\mathbb{R}$, thus the object "stochastic cosurface" is an extension of stochastic processes when the time point t are replaced by d-1-dimensional hypersurfaces c of M.

A complex K is an ordered n-tuple $K = \{c_1, \ldots, c_n\}$ with $c_i \in \sum_M$ (for $d = 1$ K is just an n-tuple of "times").

Define $m(K) \equiv \{m(c_1), \ldots, m(c_n)\}$, then we can look upon m(K) as an element in G^K.

For $d = 1$, K is just an n-tuple of "times" and m(K) is a random vector built by the process evaluated at this time points. Two stochastic cosurfaces m, $\tilde{m}$ with underlying probability measures P, $\tilde{P}$ are equivalent if their distributions are equal as measures on G^K, we shall henceforth identify equivalent stochastic cosurfaces.

We shall now introduce the concept of Markov stochastic cosurfaces and for this we need the concept of regular complexes.

We call a complex $K \equiv \{c_1,\dots,c_n\}$ regular if $c_i \cap c_j \subset \partial c_i \cap \partial c_j$ for $i \neq j$, if $\partial c_i \cap \partial c_j \neq \phi$ or else $c_i \cap c_j$ is 1-dimensional.

For any subset Λ of M we shall denote by $\sum(\Lambda)$ the σ-algebra generated by all stochastic variables $m(c)$, $c \in \sum_M$, $c \subset \Lambda$.

We say, that a subset $c_j,\dots,c_\ell$ is a splitting system for a regular complex $K = \{c_1,\dots,c_n\}$ if the complement $M - \bigcup_{i=j}^{\ell} c_j$ consists of 2 connected components $M^{\pm}$ s.t. $c_1,\dots,c_{j-1} \subset \overline{M^+}$ and $c_{\ell+1},\dots,c_n \subset \overline{M^-}$, where - means closure.

We say, that a stochastic cosurface m has the (global) Markov property with respect to the splitting system $c_j,\dots,c_\ell$ if, with $c = \bigcup_{i=j}^{\ell} c_i$:

$$E(f^+f^- | \textstyle\sum(c)) = E(f^+|\sum(c))\, E(f^-|\sum(c)),$$

for any $f^{\pm}$ such that the expectations exists and f^+ is $\sum(M^+ \cup c)$ resp. f^- is $\sum(M^- \cup c)$-measurable.

We say that m has the Markov property or is Markov whenever it has the Markov property with respect to arbitrary splitting systems.

Remark For $d = 1$ this property corresponds to the global Markov property of G-valued processes. For $d \geq 1$, $G = \mathbb{R}$ it corresponds to the global Markov property of (generalized) random fields.

We shall call Markov cosurface any stochastic cosurface which has the above Markov property: in the case $d = 2$ we also use the name of Markov (G-valued) curve integral for a Markov cosurface. We shall now see that given a generalized invariant Markov semigroup p_A on G indexed by $(M,\mathcal{B})$ (or equivalently a G-valued stochastic multiplicative measure on $(M,\mathcal{B})$) we can construct a Markov cosurface $m(c)$, $c \in \sum_M$. For this we need the concept of regular saturated complex.

We call in this way any regular complex $K = \{c_1,\dots,c_n\}$ s.t. there exists a partition $D_K = \{A_1,\dots,A_m\}$ of M into connected and simply connected closed subsets A_i of M s.t. M is the union of the A_i and if $A_i \cap A_j \neq \phi$ for some $i \neq j$, then either $A_i \cap A_j$ is d-2-dimensional or else $A_i \cap A_j$ is a piecewise smooth d-1 hypersurface which can be written as the union of some of the c_i s.t. $\bigcup_{i\neq j} A_i \cap A_j = \bigcup_{i=1} c_i$. It can be shown [33] that if G is abelian for $d > 2$ or general (not necessarily abelian) for $d = 2$, then one can construct a probability measure P on $\Gamma_{M,G}$ s.t. for any regular saturated complex $\tilde{K}$ restriction $P_{\tilde{K}}$ to $G^{\tilde{K}}$ is given by $d\, P_{\tilde{K}}(m(\tilde{K})) = \prod_{A\in D_{\tilde{K}}} p_A(\prod_{c\in\partial A\cap\tilde{K}} m(c))$, (6.1)

the product being the ordered one.

The restriction of P to an arbitrary regular complex is then given by

$$dP(m(K)) = \int dP_{\tilde{K}}(m(\tilde{K})),$$

the integration being over any regular saturated complex $\tilde{K} = (\tilde{c}_1,\dots,\tilde{c}_n)$ s.t. $m(c) = \Pi m(\tilde{c}_i)$, whenever $c = \bigcup_i \tilde{c}_i$, $c \in K$.

In fact p is determined by p_K (hence by p) as a projective limit.

The "coordinate function" m(c) with underlying probability space $(\Gamma_{M,G},P)$ is then a Markov cosurface, as easily verified see [33].

By restricting the Markov cosurface m to d-1-dimensional hypersurfaces which enclose d-dimensional regions A_c and setting $\tilde{\eta}(A_c) \equiv m(c)$ we associate to the Markov cosurface a quantity $\tilde{\eta}(A_c)$ which can easily be seen to have the properties of a stochastic G-valued multiplicative measure (here we look at A_c simply as subset of $(M,\mathcal{B})$).

Call $\mathcal{R}$ the class of all measurable sets A_c of the above form, looked upon as subsets of $(M,\mathcal{B})$.

It is not difficult to verify that $\tilde{\eta}$ is the restriction to $\mathcal{R}$ of the stochastic G-valued multiplicative measure η associated, by the results of Sect. 2, with the generalized invariant Markov semigroup p.

Let us briefly indicate the relation which exists for $d = 2$ between Markov curve integrals m and the "stochastic connections" which realize (Euclidean) gauge fields.

In analogy with (5.3) we consider the object $m(\partial dS_x)$, which is well defined by taking ∂dS_x as a (one-member) complex.

Let $F(\tilde{\omega})$ be the (singular) stochastic 2-form s.t.

$$\langle F(\tilde{\omega}), dS_x\rangle = m(\partial dS_x)$$

Formally then F is the stochastic curvature corresponding to the stochastic connection $m(\partial dS_x)$, i.e. $\int_{\partial dS_x} \tilde{\omega} = m(\partial dS_x)$, with $\tilde{\omega}$ a stochastic 1-form. By the discussion in Sect. 5, F is then the stochastic curvature form of a stochastic gauge field.

Finally let us point out the relation, in the case of general d, of our Markov cosurface with relativistic quantum fields, see [33b] for more details. Let us consider the Markov cosurfaces m, in the case where the generalized semigroup p giving η is of the special form $p_A = P_{\lambda(A)}$, with P a Markov semigroup on G and λ(A) the volume measure on the oriented Riemannian manifold M. In this case one can easily show that the finite dimensional distributions of m are invariant under the transformations induced by piecewise smooth transformations of M leaving λ and the orientation of M invariant. Moreover, if $p_t(h^{-1}) = p_t(h)$ we have the independence of m from the orientation on M.

In the case $M = \mathbb{R}^d$ one can associate to such an invariant Markov cosurface a Markov semigroup T_t acting on a Hilbert space $\mathcal{H}$ of functions. $\mathcal{H}$ is constructing from equivalence classes of real-valued bounded continuous functions (on G^{K^+}, $K^+ = \{c_1^+,\dots,c_n^+\}$, $c_i^+ \subset \{(x^1,\dots,x^d) \in \mathbb{R}^d \mid x^1 > 0\}$, with scalar product $(f^+,g^+)_{\mathcal{H}} = E(f^+(g^+ \circ T) \mid \Sigma(c_o))$, with $c_o \equiv \{x^1 = 0\}$ and T the transforms on functions induced by reflections with respect to c_o. T_t is defined by $(T_t f)(m(K)) \equiv f(m(K_t))$, with K_t the complex translated by t in the x^1-direction. T_t is then a symmetric Markov semi group on $\mathcal{H}$. By an adaptation of methods of axiomatic field theory [49] one can get then, in the case of Euclidean invariant Markov cosurfaces, a relativistic theory with Hamiltonian coinciding with the infinitesimal generator of T_t, in which the relativistic fields are associated with (closed) d-1-dimensional hypersurfaces.

7. G-valued Markov cosurfaces as limits of Gibbsian G-valued random fields

We shall here report shortly on results essentially contained in [33], which imply that the Markov cosurfaces and G-valued stochastic integrals constructed in the previous sections can be obtained as continuum limit of their discretization, which is described by discrete Gibbsian random fields. For simplicity we shall here assume $M = \mathbb{R}^d$, $\mathcal{B} = \mathcal{B}(\mathbb{R}^d)$.

We consider the discrete lattice $L_\varepsilon \equiv \varepsilon\mathbb{Z}^d$. Let γ be the subset of $\mathbb{R}^d$ corresponding to an oriented elementary cell of L_ε and let us call face F any d-1-dimensional oriented hyperface belonging to the boundary $\partial\gamma$ of γ.

For any F_1, F_2 we define $F_1 \cdot F_2$ as in Sect. 5.

For any face F define m(F) to be a G-valued cosurface.

Let U be an invariant real-valued function on G and let β be a real constant. Let Λ be any finite union of cells. We shall consider the following "Gibbs interaction" W on L_ε:

$$W_\Lambda(m) \equiv -\beta \sum_{\gamma\subset\Lambda} U(m(\partial\gamma)).$$

Let us define the probability measure μ_Λ, depending on the variables m(F) associated with faces in Λ:

$$d\mu_\Lambda(m(F), F\subset\Lambda) \equiv Z_\Lambda^{-1} \exp(W_\Lambda(m)) \prod_{F\subset\Lambda} dm(F),$$

where Z_Λ is the normalizing factor making μ_Λ into a probability measure on G^{N_Λ}, with N_Λ the number of faces in Λ. dh, for $h \in G$, stands for the Haar measure on G.

We call Gibbs state μ_ε to the interaction W any probability measure on the natural product space associated with L_ε s.t.

$$E_{\mu_\varepsilon}(f \mid \Sigma(M-\Lambda)) = E_{\mu_\Lambda}(f \mid \Sigma(M-\Lambda)),$$

for any $\Sigma(\Lambda^0)$ measurable function f for which expectations are defined, with $\Lambda^0 \subset \Lambda$. μ_ε has the Markov property in a sense corresponding to the one of Sect. 6. We call (m(F), F face of L_ε) with underlying probability measure μ_ε a <u>lattice cosurface with Gibbs distribution and interaction U.</u>

<u>Remark</u> For d = 2, G compact, $U(h) = \text{Re}\,\chi(h)$, χ a character of a unitary representation of G, we have that a lattice cosurface with Gibbs interaction describes a lattice gauge field theory (on the lattice L_ε).

It is possible to show that for a suitable choice of $\beta = \beta(\varepsilon)$ and of U, the above Gibbs measure μ_ε converges as $\varepsilon \downarrow 0$ weakly to a measure μ which can be identified with the probability measure P giving the distribution described by (6.1) of the

G-valued Markov cosurfaces associated with $\mathbb{R}^d$.

Examples:

a) $G = U(1) = \{e^{i\varphi}, \varphi \in [0,2\pi)\}$, $U(e^{i\varphi}) = \mathrm{Re}\chi(e^{i\varphi}) = \cos\varphi$, $\beta(\varepsilon) = 2^d/\varepsilon$.

In this case one gets P as the probability measure described by (6.1), with $p_A = p_{|A|}$, with Markov semigroup $p_t = e^{\frac{t}{2}\Delta}$, $t > 0$, $-\Delta$ being the Laplace-Beltrami operator on U(1).

b) $d = 2$, $G = SU(2)$, $U(h) = \sin t/\sin(t/2)$, $t \in [0,2\pi]$, with $e^{\pm it/2}$ the eigenvalues of $h \in SU(2)$, $\beta(\varepsilon) = 2^d/\varepsilon$. In this case P is as in a) with $-\Delta$ the Laplace-Beltrami operator on G.

c) $d = 2$, $G = Z^2 = \{\pm 1\}$, $U(h) = h$, $\beta(\varepsilon) = \ln\sigma + \frac{1}{\varepsilon}\ln 2$. In this case P is as in a), with p_t replaced by $p_t(h) = e^{-\frac{t}{2}\sigma^2} h+1$.

This gives some perhaps more intuitive picture of the Markov cosurfaces discussed in Sect. 6, also in relation with the generalized Markov semigroups and multiplicative stochastic measures.

Much remains of course to be done to exploit the results of this work.

In particular work on extending the results concerning the "gauge fields case" d = 2 to the case of "Higgs fields" is in preparation. On a more general line it is hoped that some of the work described above might provide new stimulation for the investigation of stochastic analysis for forms on manifolds and associated Markov fields.

Acknowledgement

It is a pleasure to thank A. Kaufmann and Dr. W. Kirsch for clarify discussions, we are grateful to the Mathematics Departments of Bochum, Marseille-Luminy and Oslo and the Zentrum für Interdisziplinäre Forschung (Project No 2, organized by Prof. Dr. L. Streit) for kind invitations which greatly facilitated this work. The partial financial support by the Norwegian Science Council (NAVF), under the program "Mathematic Seminar, Oslo" and by the Volkswagenstiftung, Forschungsprojekt BiBoS, is also gratefully acknowledged. We thank Mrs. Mischke and Richter for the skilful typing.

References

1 U. Grenander, Probabilities on Algebraic STructures, Wiley, New York (1963)

2 K. Yosida, Functional Analysis, Springer, Berlin (1965)

3 G.A. Hunt, Markoff processes and potentials: I, II, III, Ill. J. Math. 1, 44-93; 316-369 (1957); 2, 151-213 (1958)

4 R. Azencott et al., Géodésiques et diffusions en temps petit, Astérique 84-85 (1981)

5 N. Ikeda, S. Watanabe, Stochastic Differential Equations and Diffusion Processes, North Holland, Amsterdam (1981)

6 D. Elworthy, Stochastic Differential Equations on Manifolds, Cambridge Univ. Press (1982)

7 S. Albeverio, T. Arede, The relation between quantum mechanics and classical mechanics: a survey of some mathematical aspects, ZiF-Preprint, to appear in "Quantum chaos", Ed. G. Casati, Plenum Press (1985)

8 K. Ito, Stochastic differential equations on a differentiable manifold 1 Nagoya Math. J. 1, 35-47 (1950); 2, Mem. Coll. Sci. Kyoto Univ. 28, 82-85 (1953)

9 R.Gangolli, On the construction of certain diffusions on a differentiable manifold, Zeitschr. f. Wahrschienlichkeitsth. 2, 209-419 (1964)

10 H.P. McKean, jr., Stochastic Integrals, Academic Press (1969)

11 E. Jørgensen, Construction of the Brownian motion and the Ornstein Uhlenbeck process in a Riemannian manifold on the basis of the Gangolli-McKean injection scheme, Z. Wahrscheinlichkeitsth. verw. Geb. 44, 71-87 (1978)

12 T.E. Duncan, Stochastic system theory and affine Lie algebras, Proc. Berkeley, Ed. L.R. Hurt, C.F. Martin, Math. Sci. Press (1984);
T.E. Duncan, Stochastic systems in Riemannian manifolds, J. Optim. Th. and Appl. 27, 399-425 (1979)

13 M. Ibero, Integrales stochastiques multiplicatives et construction de diffusions sur un groupe de Lie, Bull. Sci. Math. 100, 175-191 (1976)

14 M. Pinsky, Mean exit times and hitting probabilities of Brownian motion in geodesics balls and tubular neighborhoods, BiBoS-Preprint 1985, Proc. BiBoS Symp. 1, Lect. Notes Maths., Springer (1985)

15 S. Albeverio, R. Høegh-Krohn, Diffusion fields, quantum fields, and fields with values in Lie groups, pp 1-98 in Stochastic Analysis and Applications, Edt. M. Pinsky, M. Dekker (1985)

16 S. Albeverio, R. Høegh-Krohn, D. Testard, Factoriality of representations of the group of paths on SU(n), J. Funct. Anal. 57, 49-55 (1984)

17 M. Fukushima, Energy forms and diffusion processes, BiBoS-Preprint 1985, Proc. BiBoS Symp. 1, Lect. Notes Maths., Springer (1985)

18 M. Pinsky, Can you feel the shape of a manifold with Brownian motion, Exp. Math. 2, 263-271 (1984)

19 P.A. Meyer, Géometrie stochastique sans larmes, Séminaire de Probabilités XV, 1979/80, Lect. Notes Maths. 850, Springer (1981)

20 D. Williams, Ed., Stochastic Integrals, Proc., LMS Durham Symp., 1980, Lect. Notes Maths. 851, Springer (1981)

21 J.M. Bismut, The Atiyah-Singer theorems for classical elliptic operators: a probabilistic approach I. The index theorem, J. Funct. Anal. 57, 56-99 (1984)

22 P. Malliavin, Géometrie différenticlle stochastique, Presses Univ. Montreal (1978)

23 G.A. Hunt, Semi-groups of measures on Lie groups, Trans. Am. Math. Soc. 81, 264-293 (1956)

24 A. Hulanicki, A class of convolution semi-groups of measures on a Lie group, in Probability Theory on Vector Spaces II, Ed. A. Wron, Lect. Notes Maths. 828, Springer (1980)

25 Ph. Feinsilver, Processes with independent increments on a Lie group, Trans. Am. Math. Soc. 242, 73-121 (1978)

26 R. Cairoli, J. Walsh, Stochastic integrals in the plane, Acta Math. 134, 111-183 (1975)

27 E. Wong, M. Zakai, Multiparameter martingale differential forms, Preprint (1985)

28 a) M. Dozzi, 2-parameter harnesses and the Wiener process, Z. Wahrscheinlichkeitsth. verw. Geb. 56, 507-514 (1981)

b) D. Nualart, M. Sanz, Malliavin calculus for two-parameter Wiener functionals, Barcelona-Preprint (1985)

c) H. Korezlioglu et al, Edts., Processus aléatoires à deux index, Proc., Lect. Notes Maths. 863, Springer, Berlin (1981)

29 S. Albeverio, R. Høegh-Krohn, Aremark on dynamical semigroups in terms of diffusion processes, BiBoS-Preprint (1985), to appear in Proc. Heidelberg Conf. Quantum Probability, Ed. L. Accardi et al.

30 S. Albeverio, R. Høegh-Krohn, J. Marion, D. Testard, book in preparation

31 G.F. De Angelis, D. De Falco, G. Di Genova, Random fields on Riemannian manifolds, A constructive approach, Salerno-Preprint 1984

32 Z. Haba, Instantons with noise. I. Equations for two-dimensional models, BiBoS-Preprint (1985)

33 a) S. Albeverio, R. Høegh-Krohn, H. Holden, Markov cosurfaces and gauge fields, Acta Phys. Austr. XXVI, 211-231 (1984)

b) S. Albeverio, R. Høegh-Krohn, H. Holden, Some models of Markov fields and quantum fields, trough group-valued cosurfaces, in preparation

c) S. Albeverio, R. Høegh-Krohn, H. Holden, Markov processes on infinite dimensional spaces, Markov fields and Markov cosurfaces, BiBoS-Preprint, to appear in Proc. Bremen, Conf. 1983, Ed. L. Arnold, P. Kotelenez, D. Reidel (1985)

34 K. Ito, Isotropic random current, Proc. Third Berkeley Symp., Univ. Berkeley

35 M. Baldo, Brownian motion in group manifolds: application to spin relaxation on molecules, Physica 114A, 88-94 (1982)

36 D. Wehn, Probabilities on Lie groups, Proc. Nat. Ac. Sci 48, 791-795 (1962)

37 R.M. Dowell, Differentiable approximations to Brownian motion on manifolds, Ph. Thesis, Warwick (1980)

38 D.W. Stroock, S.R.S. Varadhan, Limit theorems for random walks on Lie groups, Sankhya Ser. A(3) 35, 277-294 (1973)

39 S. Albeverio, R. Høegh-Krohn, H. Holden, Stochastic multiplicative measures, generalized Markov semigroups and group valued stochastic processes, BiBoS-Preprint, in preparation

40 A.V. Skorohod, Studies in the theory of random processes, Dover, New York (1982)

41 S.G. Bobkov, Variations of random processes with independent increments, J. Sov. Math. 27, 3167-3340 (1984)

42 I.M. Gelfand, N.Ya. Vilenkin, Generalized functions, Vol. IV; Academic Press (1964)

43 Yu.V. Prohorov, Yu.A. Rozanov, Probability Theory, Springer, Berlin (1969)

44 D.A. Dawson, Generalized stochastic integrals and equations, Trans. Am. Math. Soc. 147, 473-506 (1970)

45 J. Kerstan, K. Matthes, J. Mecke, Unbegrenzt teilbare Punktprozesse, Akademie Verlag, Berlin (1974)

46 M.M. Rao, Local functionals and generalized random fields with independent values, Th. Prob. and Appl. 16, 457-473 (1971)(transl.)

47 A. Tortrat, Sur le support deslois indéfiniment divisibles dans les espaces vectoriels localement convexes, Ann. I.H. Poincaré B 13, 27-43 (1977)

48 S. Albeverio, R. Høegh-Krohn, H. Holden, W. Kirsch, Higgs fields in two dimensions multiplicative integrals, polymer representation, papers in preparation

49 E. Seiler, Gauge theories as a problem of constructive quantum fiels theory and statistical mechanics, Lect. Notes Phys. 159, Springer, Berlin (1982)

50 R.V. Kohn, Integration over stochastic simplices, Warwick M. Sc. Diss. (1975)

51 J. Potthoff, Stochastic path-ordered exponentials , J.Math.Phys.25,52-56 (1984)

EXISTENCE AND SAMPLE PATH PROPERTIES OF THE DIFFUSIONS IN NELSON'S STOCHASTIC MECHANICS

Eric A. Carlen

Department of Mathematics, M.I.T., Cambridge MA, 02139, USA

INTRODUCTORY REMARKS

Nelson's stochastic mechanics offers a description of quantum phenomena in terms of diffusions instead of wave functions. It is not yet a complete theory, either physically or mathematically; but this has a positive aspect: stochastic mechanics is still the source of many interesting questions. Here, I will discuss three questions arising in this subject.

Before I even tell you what the questions are, I will spend a few minutes discussing stochastic mechanics in a very simple context. This should make the considerations that follow more concrete, and it is probably the most convenient way to fix the necessary notation.

A DESCRIPTION OF STOCHASTIC MECHANICS

For the purpose of introduction, we will consider stochastic mechanics in the context of a single particle moving in $\mathbb{R}^3$ under the influence of a potential $V(x)$. Mechanics - whatever adjective one places before it - consists of two parts: kinematics and dynamics. In the context at hand, the kinematical part of our theory is to explain what we mean by "moving", and the dynamical part is to explain what we mean by "influence of a potential". Some mathematical details will be glossed over in the introduction, but when we get to the actual theorems, we will be precise.

The kinematical part of stochastic mechanics is that the motion of our particle is to be given by a Markovian diffusion process

$$t \mapsto \xi(t) \tag{1}$$

in $\mathbb{R}^3$ which solves a stochastic differential equation of the form

$$d\xi(t) = b(\xi(t), t)dt + \left(\frac{\hbar}{m}\right)^{1/2} dw(t) \tag{2}$$

where $t \mapsto w(t)$ is a standard Brownian motion, and $b: \mathbb{R}^3 \times \mathbb{R} \to \mathbb{R}^3$ is some time dependent vector field called the drift field.

While the first term on the right hand side of (2) has a rather general form, the second term is particularly simple. The decision to restrict our attention to diffusions of this form is motivated by the following physical considerations: The position of the particle at time t is taken to be a random variable in the first place on account of "quantum fluctuations". I would like to be more specific about the physical meaning of this term, but I can't. Perhaps my predicament is like that of a nineteenth century physicist who wants to explain "Brownian fluctuations", and who doesn't know about molecules.

But even without a clear understanding of what quantum fluctuations are, it is reasonable to assume that they are the manifestation of some isotropic, translation invariant phenomenon. That assumption made, it follows that the noise term in our stochastic differential equation must be a constant times the increment of a Brownian motion. Since the mean of $|w(t) - w(s)|^2$ is $|t-s|$, it follows that $w(t)$ has units $(\text{time})^{1/2}$. Since $\xi(t)$ is to have units of distance, it follows that the constant must have units $(\text{distance})/(\text{time})^{1/2}$. $(\hbar/m)^{1/2}$ is such a constant; this choice is fully motivated by the dynamical considerations to follow.

The dynamical part of stochastic mechanics is given by the Guerra-Morato variational principle [2]. We will not give details here, as Guerra himself has spoken on it at this conference, but will just remind you that it is a beautiful and direct translation of the Lagrangean variational principle of classical mechanics into the kinematical context just discussed.

A theorem of Guerra and Morato then asserts that a diffusion is critical for their variational principle precisely when there is a solution $\psi(x,t)$ of the Schroedinger equation

$$i\hbar \frac{\partial}{\partial t}\psi(x,t) = \left(-\frac{\hbar^2}{2m}\Delta + V(x)\right)\psi(x,t) \tag{3}$$

so that the drift field $b(x,t)$ of the diffusion is given by

$$b(x,t) = \frac{\hbar}{m}\left(\mathrm{Re}\,\frac{\nabla\psi}{\psi}(x,t) + \mathrm{Im}\,\frac{\nabla\psi}{\psi}(x,t)\right) \tag{4}$$

and so that the probability that $\xi(t)$ is in a measurable set A is given by

$$\Pr\{\xi(t) \in A\} = \int_A |\psi(x,t)|^2 dx \tag{5}$$

This last equation can be expressed by saying that $\xi(t)$ has a density $\rho(x,t)$ given by

$$\rho(x,t) = |\psi(x,t)|^2 \tag{6}$$

Note that it is the same potential which appears in (3) and which governs the motion of the diffusing particle. The choice of $(\hbar/m)^{1/2}$ as the constant appearing in (2) is responsible for the fact that $\hbar$ and m enter the Schroedinger equation in the usual way. This said we put $\hbar = m = 1$ in the rest of our talk.

Equation (5) is particularly interesting; it says that stochastic mechanics and ordinary quantum mechanics make the same predictions for the same position measurement experiments. The two descriptions cannot be distinguished experimentally.

The motion of a single spinless particle does not at all exhaust the scope of stochastic mechanics; moreover, our discussion of even this simple case has skipped over much. For a comprehensive discussion, see the forthcoming book "Quantum Fluctuations" [1] by Ed Nelson.

The intriguing relation between stochastic mechanics and ordinary quantum mechanics which we have just described raises a host of questions, and there has been much discussion of stochastic mechanics in the recent literature. See, for instance, the bibliography to [1]. Here we will focus on the following three questions:

(I) Do the diffusions of stochastic mechanics really exist? A glance at the formula for $b(x,t)$ in terms of $\psi(x,t)$ is enough to see that the drift fields are extremely singular objects in general - much too singular to permit, say, direct application of the Girsanov formula.

The first theorem we will report on here provides a positive answer to this question. Next we ask:

(II) Do the diffusions of stochastic mechanics have physically reasonable behavior pathwise?

Since the sample paths of these diffusions are supposed to be the possible actual particle motions, this question is basic to stochastic mechanics. Nonetheless, there are very few results, even now, bearing on this question.

This question must be answered anew in each physical context, so our answer here cannot be as complete as the answer we will give to the first question. Here we will dicuss potential scattering in stochastic mechanics, and in this context present a theorem which provides a pleasing positive answer.

Knowing that the sample paths of these diffusions have the "right" behavior, we go on to ask a very optimistic question:

(III) Can one use direct probabilistic analysis of the sample paths of these diffusions to study the behavior of quantum systems?

This is perhaps the most exciting question of the three, but unfortunately I do not yet have a theorem here to report. By the end of the talk, I will at least be able to rephrase this question as a rather specific mathematical problem.

There are of course many other interesting questions one can ask, and much interesting work has been done by other people. For most of this other work, I unfortunately only have time to refer you to [1] and its bibloography.

DIFFUSION THEORY

In this section of the talk, I will quickly review diffusion theory from a time symmetric point of view; and, in the process, establish enough notation and terminology to treat the questions of the last section.

Let Ω_c be the space of continuous functions $\omega : \mathbb{R} \longrightarrow \mathbb{R}^n$ given the topology of uniform convergence on compacts; we call this space trajectory space. There is a distinguished class of functions on Ω_c : the t-configuration function $\xi(t)$ is defined by

$$\xi(t) : \omega \longmapsto \omega(t) \equiv \xi(t,\omega) \tag{7}$$

It is a theorem that the Borel field $\mathcal{B}$ on Ω_c is generated by the configuration functions:

$$\mathcal{B} = \sigma\{\xi(t) \mid t \in \mathbb{R}\} \tag{8}$$

There are certain sub fields of $\mathcal{B}$ of special interest: the past, present, and future are given by

$$\mathcal{P}_t = \sigma\{\xi(s) \mid s \le t\}, \quad \mathcal{N}_t = \sigma\{\xi(t)\}, \quad \mathcal{F}_t = \sigma\{\xi(u) \mid u \ge t\} \tag{9}$$

Now suppose we are given a Borel probability measure Pr on Ω_c. Then Borel functions f become random variables, and we denote their expectations and conditional expectations - provided they are integrable - in the usual way, except perhaps that we write

$$E_t f = E\{f \mid \mathcal{N}_t\} \tag{10}$$

When trajectory space is made into a probability space in this way, $t \mapsto \xi(t)$ becomes a stochastic process which we shall call the <u>configuration process</u>.

At last we come to the definition of a diffusion:

<u>DEFINITION</u>: We will say that under Pr, the configuration process is a diffusion with forward generator $\mathcal{D}$ and backwaed generator $\mathcal{D}_*$ in case

(i) Under Pr, $t \mapsto \xi(t)$ is Markovian.

(ii) $\mathcal{D}$ and $-\mathcal{D}_*$ are (possibly time dependent) elliptic operators of the form

$$\mathcal{D} = \frac{1}{2} a^{ij} \frac{\partial^2}{\partial x^i \partial x^j} + b^i \frac{\partial}{\partial x^i} \qquad \mathcal{D}_* = -\frac{1}{2} a^{ij} \frac{\partial^2}{\partial x^i \partial x^j} + b_*^i \frac{\partial}{\partial x^i} \tag{11}$$

(iii) For any $f \in C_0^\infty(\mathbb{R}^n)$ and $T \in \mathbb{R}$:

$$f(\xi(t)) - f(\xi(T)) - \int_T^t (\mathcal{D} f)(\xi(\tau))\, d\tau \tag{12}$$

is a $\mathcal{P}_t$ martingale on $[T,\infty)$ and

$$f(\xi(T)) - f(\xi(t)) - \int_t^T (\mathcal{D}_* f)(\xi(\tau))\, d\tau \tag{13}$$

is a $\mathcal{F}_t$ martingale on $(-\infty,T]$.

Note that the definition doesn't include any explicit regularity conditions on the operators $\mathfrak{D}$ and $\mathfrak{D}_*$. This is intentional, but when we are dealing with coefficients as singular as those discussed in the first section, we must explain what we mean by elliptic in (ii). Let $\rho(dx,t)$ be the image of Pr under $\xi(t)$; this is a measure on $\mathbb{R}^n$. When we say that $\mathfrak{D}$ is elliptic, we mean that almost everywhere with respect to this measure, $|b(x,t)| < \infty$, and $a^{ij}(x,t)$ is a matrix of strictly positive type. Also, it is part of the definition that the integrals in (iii) exist pathwise.

This definition of diffusion is clearly time symmetric, though it may look like we have contrived this by appending refernces to $\mathfrak{D}_*$ and backward martingales to a version of the classical Stroock-Varadhan martingale problem definition. However, it is not hard to see that if the coefficients of $\mathfrak{D}$ are sufficiently regular, and if the conditions in our definition refering to $\mathfrak{D}$ are satisfied, then there exists a vectorfield $b_*(x,t)$ on $\mathbb{R}^n \times \mathbb{R}$ so that if $\mathfrak{D}_*$ is defined by (11), then the rest of the definition holds. In fact, in this case the measure $\rho(dx,t)$ will be of the form $\rho(x,t)dx$ where $\rho(x,t)$ is smooth by classical parabolic regularity theorems; and moreover $\rho(x,t)$ will be everywhere free of zeros by the strong maximum principle. Then we have [1] the following explicit formula relating b, b_*, and ρ:

$$b_*^i = b^i - \frac{1}{\rho}\frac{\partial}{\partial x^j}(a^{ij}\rho) \tag{14}$$

Trivially modifying our definition so that we can work on a finite time interval, an example is Brownian motion on [0,T] with the initial density $(2\pi)^{-n/2}\exp(-x^2/2)$ Then:

$$a^{ij}(x,t) = \delta^{ij} \qquad b^i(x,t) = 0 \qquad b_*^i(x,t) = -x^i/(1+t) \tag{15}$$

As indicated in the introduction, we are interested in the case where the fluctuations are isotropic and translation invariant so that the noise term in our stochastic differential equation is a Brownian motion. In terms of the generator, it is equivalent to require that

$$a^{ij}(x,t) = \delta^{ij} \quad \forall\, x,t \tag{16}$$

We henceforth restrict attention to this case, so our generators are of the form

$$\mathcal{G} = \frac{1}{2}\Delta + b\cdot\nabla \qquad \mathcal{G}_* = -\frac{1}{2}\Delta + b_*\cdot\nabla \tag{17}$$

We will find the odd and even combinations of b and b_* particalarly useful; following Nelson we define the osmotic velocity u(x,t) by

$$u(x,t) = \frac{1}{2}\left(b(x,t) - b_*(x,t)\right) \tag{18}$$

and the current velocity v(x,t) by

$$v(x,t) = \frac{1}{2}\left(b(x,t) + b_*(x,t)\right) \tag{19}$$

It is then a theorem [1] that u, v, and ρ are related by

$$u = \frac{1}{2}\nabla \log \rho \qquad \frac{\partial}{\partial t}\rho = -\nabla\cdot(v\rho) \tag{20}$$

This last equation is called the continuity equation, and we use it repeatedly in what follows.

CONSTRUCTING DIFFUSIONS

The components of b, b_*, and so forth are often called the "infinitessimal characteristics" of a diffusion. We have discussed some relations among the infinitessimal characteristics of a given diffusion. Now we turn to the problem of constructing a diffusion which has a given set of infinitessimal characteristics. We will approach this problem from a rather analytic point of view, but one which leads to a method which works for even the singular coeeficients that are our main interest. The hard part of the work in this case to solve a certain partial differential equation, but before plunging into this, I will briefly describe the method in the case of smooth coefficients where the analysis is easy.

Fix a compact interval [0,T] and a smooth bounded vector field $b(x,t)$ on $\mathbb{R}^n\times[0,T]$. Suppose we already have a measure Pr making $t \mapsto \xi(t)$ a diffusion on [0,T] with backward generator $-\frac{1}{2}\Delta+b_*\cdot\nabla$. Let $f(x,t)$ be a smooth function on $\mathbb{R}^n\times[0,T]$. (Smooth means possesing bouded derivatives of all orders.) Then the process

$$t \mapsto f(\xi(t),t) \tag{21}$$

is a backward martingale precisely when f satisfies the <u>backward martingale equation</u>:

$$\frac{\partial}{\partial t}f(x,t) = \left(\tfrac{1}{2}\Delta - b_*\cdot\nabla\right)f(x,t) \tag{22}$$

As the name implies, this follows almost directly from (13). This equation is parabolic, and since b_* is smooth, classical theorems gaurantee the existence, uniqueness, and regularity of a Markovian transition function $\tilde{p}_*(y,t;x,s)$ - the fundamental solution - which generates its solutions according to

$$f(y,t) = \int \tilde{p}_*(y,t;x,s)f(x,s)\,dx \tag{23}$$

But since $t \mapsto f(\xi(t),t)$ is a backward martingale, it is also true that

$$f(\xi(t),t) = E_t f(\xi(s),s) \tag{24}$$

Now, future conditional expectations can be computed in terms of the backward transition function of our process. That is, if

$$Pr\{\xi(s)\in dx \mid \xi(t)=y\} = p_*(y,t;x,s)\,dx \tag{25}$$

then

$$f(\xi(t),t) = \int p_*(\xi(t),t;x,s)\,f(x,s)\,dx \tag{26}$$

Comparing equations, it is evident that

$$p_*(y,t;x,s) = \tilde{p}_*(y,t;x,s) \tag{27}$$

So, to find the backward transition function of Pr, what we have to do is to solve the corresponding backward martingale equation for its transition function. Moreover, the uniqueness of bounded solutions of the backward martingale equation immediately implies that Pr is the unique measure under which $\xi(T)$ has density $\rho(x,T)$ and $t\mapsto\xi(t)$ is a diffusion with backward generator $-\frac{1}{2}\Delta + b_*\cdot\nabla$.

Now we drop the assumption that we already have Pr; and instead we just assume that we have u, v, and ρ. We want to contsruct a diffusion having them as its osmotic velocity, current velocity, and density respectively. In the next section we will spell out conditions under which it is possible to do this, but the considerations of this section already provide a strategy: We define b_* to be v−u, and we solve for the fundamental solution $p_*(y,t;x,s)$ of the corresponding backwards martingale equation

$$\frac{\partial}{\partial t}f(x,t) = \left(\tfrac{1}{2}\Delta - (v-u)\cdot\nabla\right)f(x,t) \tag{28}$$

Then with $p_*(y,t;x,s)$ and $\rho(y,t)$ in hand, we construct a measure Pr on in a famililar fashion. In fact, we don't even need $p_*(y,t;x,s)$; we only need the operator $P_{t,A}$ given by $P_{t,A}f(y) = \int p_*(y,t;x,s)f(x)\,dx$. Having produced this measure, it remains to check that under it, $t \mapsto \xi(t)$ is indeed a diffusion with the right coefficients. This last step will actually require some work since we will only have produced a weak fundamental solution; nonetheless, we will produce an honest diffusion.

SOLVING THE BACKWARD MARTINGALE EQUATION

We begin by spelling out the conditions under which we can work. We suppose that we are given a pair (v,ρ) where v is a time dependent jointly measureable vector field, and ρ is a time dependent jointly measurable probability density whose distributional gradient has a version $\nabla\rho$ which is a jointly measurable function. We then define:

$$u(x,t) = \frac{1}{2}\frac{\nabla\rho(x,t)}{\rho(x,t)} \qquad (\equiv 0 \text{ if } \rho = 0) \tag{29}$$

We furthermore suppose that the following two conditions are satisfied:

The <u>Finite Action Condition</u>: On any finite interval $[S,T]$

$$\int_S^T \int (u^2+v^2)\,\rho(x,t)\,dx\,dt < \infty \tag{30}$$

The <u>Compatability Condition</u>: For all f C $(R$ $)$ and all finite intervals $[S,T]$

$$\int f(x)\rho(x,T)\,dx - \int f(x)\rho(x,S)\,dx = \int_S^T \int (v\cdot\nabla f)\,\rho(x,t)\,dx\,dt \tag{31}$$

This last condition requires that an integrated form of the continuity equation (20) hold so that the pair (v,ρ) is compatible. Francesco Guerra has emphasized to me the naturalness of writing (20) in the integrated form above.

Now fix a finite time interval [S,T]. The backward martingale equation initial value problem for the pair (v,ρ) then is

$$\frac{\partial}{\partial t} f(x,t) = \left(\tfrac{1}{2}\Delta - (v-u)\cdot\nabla\right) f(x,t), \quad f(x,S) = f_S(x) \tag{32}$$

I will now sketch the proof of a theorem which asserts the existence of solutions to this equation. The solutions will be weak solutions, having in general only one spatial derivative and half a time derivative; but they will posess sufficient regularity for our purpose. Part of the point here is to specify just what is required in the way of regularity.

As the first step consider a sequence of smooth bounded $\mathbb{R}^n$ valued functions $b_*(x,t)$ on $\mathbb{R}^n X[S,T]$. These are to approximate the singular function $b_* = (v-u)$. We will choose the sequence so that

$$\lim_{i\to\infty} \int_S^T \int |b_*^i - b_*|^2 \rho(x,t)\, dx\, dt = 0 \tag{33}$$

and it is here that the finite action condition first comes in.

Now fix an initial condition f_S ; at this time we require some smoothness of f_S , so take f in $C_0^\infty(\mathbb{R}^n)$.

By $f^i(x,t)$ we mean the solution to the initial value problem

$$\frac{\partial}{\partial t} f^i(x,t) = \left(\tfrac{1}{2}\Delta - b_*^i\cdot\nabla\right) f^i(x,t), \quad f^i(x,S) = f_S(x) \tag{34}$$

This is a regular parabolic initial value problem to which classical existence, uniqueness, and regularity theorems apply. In particular, we have the maximum principle for solutions of these equations:

$$\min_x f_S(x) \le f^i(y,t) \le \max_x f_S(x) \quad \forall\ (y,t) \in \mathbb{R}^n \times [S,T] \tag{35}$$

We will make repeated use of this fact.

Now in some sense it is true that our sequence of regular equations approximates the singular equation we are interested in. We will show that the solutions of our regular equations converge to a solution of the singular equation in an L^2 sense.

<u>Theorem 1</u> Let f_S, b_*^i and f^i be as above. Then

$$\lim_{i,j\to\infty} \int |f^i - f^j|^2 \rho(x,t)\,dx = 0 \tag{36}$$

converging uniformly in t [s,T] and:

$$\lim_{i,j\to\infty} \int_s^T \int |\nabla f^i - \nabla f^j|^2 \rho(x,t)\,dx\,dt = 0 \tag{37}$$

<u>Sketch of Proof</u>: These statements are the consequence of three inequalities which are derived by making repeated use of integration by parts, the Schwarz inequality and the maximum principle. The two integrations by parts which we use (over and over again) are the following:

(1) For all $f,g \in C_2^b(\mathbb{R}^n)$ and a.e. t in [S,T]

$$\int \nabla f \cdot \nabla g\, \rho(x,t)\,dx = -2 \int f\left(\left(\tfrac{1}{2}\Delta + u\cdot\nabla\right) g\right) \rho(x,t)\,dx \tag{38}$$

This follows from $u = \frac{1}{2}\nabla \log \rho$

(2) For all $g \in C_b^{2,1}(\mathbb{R}^n \times [S,T])$

$$\|g(\cdot,t)\|_t^2 - \|g(\cdot,s)\|_s^2 = \int_s^t \int g\left(\frac{\partial g}{\partial t} + v\cdot\nabla g\right)\rho(x,\tau)\,dx\,d\tau \tag{39}$$

where $\|\cdot\|_t$ is the norm on the Hilbert space $L^2(\rho(x,t)dx)$ which we henceforth denote by $\mathcal{H}_t$. We will often write $(\cdot,\cdot)_t$ for the inner product on $\mathcal{H}_t$.

As an example, just to give you the flavor, I'll derive the simplest of the three inequalities. In the proof of this theorem we need a uniform bound on $\int_S^T\int |\nabla f^i|^2 \rho(x,t)\,dx\,dt$.

Computing, this equals:

$$-2\int_S^T\int f^i\left(\left(\tfrac{1}{2}\Delta + u\cdot\nabla\right)f^i\right)\rho\,dx\,dt = -2\int_S^T\int f^i\left(\left(\tfrac{\partial}{\partial t} + b_*^i\cdot\nabla + u\cdot\nabla\right)f^i\right)\rho\,dx\,dt$$

$$= -2\int_S^T\int f^i\left(\left(-v\cdot\nabla + b_*^i\cdot\nabla + u\cdot\nabla\right)f^i\right)\rho\,dx\,dt + \|f^i\|_S^2 - \|f^i\|_T^2$$

$$\leq 2\left(\max_x |f_S(x)|\right)\left(\int_S^T\int |\nabla f^i|^2\rho\,dx\,dt\right)^{1/2}\left(\int_S^T\int |b_*^i - (v-u)|^2\rho\,dx\,dt\right)^{1/2} + \left(\max_x |f_S(x)|\right)^2$$

Now one uses the quadratic formula to deduce

$$\left(\int_S^T\int |\nabla f^i|^2\rho(x,t)\,dx\,dt\right)^{1/2} \leq$$

$$\left(\max_x |f_S(x)|\right)\left\{\left(\int_S^T\int |b_*^i - (v-u)|^2\rho\,dx\,dt\right)^{1/2} + \left(\int_S^T\int |b_*^i - (v-u)|^2\rho\,dx\,dt + 2\right)^{1/2}\right\} \tag{40}$$

The other two inequalities require more involved derivations, and indeed their final forms are complicated enough that I won't even write them down, but the approach is similar. This concludes the sketch; for the details see [3].

By the above theorem,

$$\lim_{i\to\infty} f^i(\cdot,t) \equiv f(\cdot,t) \text{ exists in } \mathcal{H}_t \;\forall\, t\in[S,T] \tag{41}$$

We will write f(x,t) to denote this limit as a funtion of x and t. Now note that $\{g:\ \min_x f \leq g(x) \leq \max_x f \text{ a.e. } x\}$ is closed in $\mathcal{H}_t$. It follows immediately that

$$\min_x f_S(x) \leq f(y,t) \leq \max_x f_S(x) \quad \text{a.e. } y\in\mathbb{R}^n,\ \forall\, t\in[S,T] \tag{42}$$

and so we have the (weak) maximum principle for our singular equation.

The next step is to remove the restrictions on f_S. To do this we prove:

Theorem 2 Let f, f_s be as above. Then

$$\int_s^t \int |\nabla f(x,\tau)|^2 \rho(x,\tau)\,dx\,d\tau + \|f(\cdot,t)\|_t^2 = \|f(\cdot,s)\|_s^2 \tag{43}$$

Furthermore, $\|f^i(\cdot,t)\|_t^2 \leq k_c^i + \|f^i(\cdot,s)\|_s^2$ where K_c^i is a constant depending only on b_*^i and $c \equiv \max_x |f_s|$ with the property that $\lim_{i\to\infty} K_c^i = 0$.

Sketch of Proof Using our second integration by parts formula,

$$\|f^i(\cdot,t)\|_t^2 - \|f^i(\cdot,s)\|_s^2 = \int_s^t \frac{d}{d\tau}\|f(\cdot,\tau)\|_\tau^2 d\tau = 2\int_s^t \left(\left(f^i, \frac{\partial f^i}{\partial}\right)_\tau + (v\cdot\nabla f^i, f^i)_\tau\right) d\tau \tag{44}$$

Now substitute $(\frac{1}{2}\Delta - b_*^i\cdot\nabla)f^i$ for $\frac{\partial f^i}{\partial t}$, use the first integration by parts formula, collect terms, and take $i \to \infty$.

This result implies that $f_s \mapsto f(\cdot,t)$ is a contraction from $C(\mathbb{R}^n)$ to $\mathcal{H}_t$ for all t in $[S,T]$. Let $P_{t,S}$ be its extension to all of $\mathcal{H}_S$ by continuity.

Nothing is special about starting at S; carrying out the same construction for all initial times s in $[S,T]$ we produce a family of contractions

$$P_{t,s} : \mathcal{H}_s \to \mathcal{H}_t \tag{45}$$

where each $P_{t,s}$ is Markovian by (42). (That is $f \geq 0$ $P_{t,s} f \geq 0$ and $P_{t,s} 1 = 1$.)

The next thing to prove is that taken together, this family of Markovian operators forms a Markovian propagator.

Theorem 3 $P_{u,s} = P_{u,t} P_{t,s}$ for all $u \geq t \geq s$.

<u>Sketch of Proof</u> We know this holds for the corresponding "approximate propagators" $P^i_{t,\Delta}$ defined in terms of the approximate equations; there it follows from uniqueness. Not having a uniqueness result for our singular equation, but having strong information on how $P^i_{t,\Delta}$ approximates $P_{t,\Delta}$ we expand $\|P_{u,\Delta}f - P_{u,t}\,P_{t,\Delta}f\|$ by adding and subtracting $P^i_{t,\Delta}$ terms and apply an $\varepsilon/5$ argument.

We have now produced a Markovian propagator which generates the "solutions" of our backward martingale equation. Note that at each time t we work with a different Hilbert space, but each contains all of the bounded measurable functions, and it is these that we are really interested in.

It only remains to determine the sense in which we have solved our equation. We have the following result:

<u>Theorem 4</u> Let $f \in \mathcal{H}_\Delta$ and $g \in C^2_b(\mathbb{R}^n)$. Then:

$$(P_{t,\Delta}f, g)_t - (f, g)_\Delta = \int_\Delta^t \left(P_{\tau,\Delta}f, \left(\tfrac{1}{2}\Delta + b\cdot\nabla\right)g\right)_\tau d\tau \tag{47}$$

<u>Remarks</u>: The proof of this theorem proceeds in much the same way as the proof of the second theorem. Note that no derivatives operate on f, and we only require f to be in $\mathcal{H}_\Delta$. Finally, its easy to see that if $P_{t,s}f_s$ were smooth in x and t, then (47) would imply that $P_{t,s}f(x)$ satisfies (34).

There are many more questions one could ask about this equation, but this is all we will need to produce our diffusion.

<u>CONSTRUCTING AND IDENTIFYING THE MEASURE</u>

We are now ready to construct our diffusion. We will first construct a probability measure Pr on a large space Ω, path space, which contains Ω_c as a measurable subset. We then show that in fact $Pr(\Omega_c) = 1$, and that under $Pr|_{\Omega_c}$, also just denoted Pr, the configuration process is a difussion with the desired generators and density.

Now for the first step. Let $\dot{\mathbb{R}}^n$ denote the one point compactification of $\mathbb{R}^n$, and let $\Omega = (\dot{\mathbb{R}}^n)^{\mathbb{R}}$ with the product topology. This makes it a compact Hausdorf space, and by a result of Nelson, Ω_c is a Borel subset. Note that the Baire field of Ω is given by $\sigma\{\xi(t): t \in R\}$. Here, $\xi(t)$ denotes the obvious extension of the configuration function on Ω_c to Ω. We also write $\mathcal{P}_t$, $\mathcal{N}_t$, and $\mathcal{F}_t$ for the obvious subalgebras of $\mathcal{B}(\Omega)$.

Next consider $\mathcal{C}_{\Delta p}(\Omega)$, the set of all functions F on Ω which can be written as a finite sum of finite products of functions of the form $f(\xi(t,\cdot))$ for some f in $C^\infty(\dot{\mathbb{R}}^n)$. Now since $\mathcal{C}_{\Delta p}(\Omega)$ is an algebra containing 1 and separating points, it is uniformly dense in $\mathcal{C}(\Omega)$.

Next, we identify each f in $C^\infty(\dot{\mathbb{R}}^n)$ with a multiplication operator on $\mathcal{H}_t$ in the usual way. We then define a linear functional Pr: $\mathcal{C}_{\Delta p}(\Omega) \to \mathbb{R}$ by putting

$$Pr(F) = \left(1, P_{T,t_n} f_n P_{t_n,t_{n-1}} f_{n-1} \cdots P_{t_2,t_1} f_1\right)_T \quad (t_1 < \cdots < t_n < T) \tag{48}$$

for F of the form $F(\omega) = \prod_{i=1}^{n} f_i(\xi(t_i,\omega))$ and taking the linear extension. Since the $P_{t,\Delta}$ are a Markovian propagator, Pr is a well defined positive linear functional on $\mathcal{C}_{\Delta p}(\Omega)$ with Pr1 = 1. It is therefore continuous, and we again let Pr denote its extension to all of $\mathcal{C}(\Omega)$. Then by the Riesz-Markov theorem, there is a unique regular Borel probability measure, again denoted Pr, so that

$$Pr(F) = \int_\Omega F(\omega)\, Pr(d\omega) \tag{49}$$

This method of constructing measures on function space is due to Kakutani and was developed into a method by Nelson [4].

The second part of our task is to show this measure provides the solution of our problem. This will be easy once we have proved the next theorem.

Theorem 5 $Pr(\Omega_c) = 1$

Sketch of Proof We must work with some care because our transition function is only almost everywhere defined and of course because of our singular coefficients. First we establish the following lemma:

Lemma Let K be a compact set in $\mathbb{R}^n$, and fix a finite time interval $[S,T]$. Then

$$Pr\{\xi(s)\in K,\ |\xi(t)-\xi(s)| > \varepsilon\} = o(t-s) \quad \text{uniformly} \tag{50}$$

Proof Let $B_y(\varepsilon) = \{x: |x-y| \leq \varepsilon\}$. Fix smooth functions F and G on $\mathbb{R}^n$ so that

$$\begin{aligned} &F \equiv 1 \text{ on } |x| \leq \tfrac{1}{2}, \quad F \equiv 0 \text{ on } |x| \geq \tfrac{3}{4}, \quad 0 \leq F \leq 1 \\ &G \equiv 1 \text{ on } |x| \geq 1, \quad G \equiv 0 \text{ on } |x| \leq \tfrac{3}{4}, \quad 0 \leq G \leq 1 \end{aligned} \tag{51}$$

Fix $y \in \mathbb{R}^n$ and $\varepsilon > 0$. Put $f(x) = F((1/\varepsilon)(x-y))$ and also $g(x) = G((1/\varepsilon)\times(x-y))$. Then by the definition of Pr,

$$Pr\{\xi(s)\in B_y(\tfrac{\varepsilon}{2}),\ \xi(t)\in B_y(\varepsilon)^c\} \leq E\big(f(\xi(s))\,g(\xi(t))\big) = (P_{t,s}f,g)_t \tag{52}$$

By our last result on the propagator,

$$(P_{t,s}f,g)_t = \int_s^t \big(P_{\tau,s}f,\ (\tfrac{1}{2}\Delta + b\cdot\nabla)g\big)_\tau\, d\tau \tag{53}$$

since $(f,g)_s = 0$. Now the fact that both directions of time enter our theory on an equal footing becomes important to us. Choosing the other direction of time to work with, we could have solved the forward martingale equation

$$\frac{\partial}{\partial t} f(x,t) = \left(-\tfrac{1}{2}\Delta - (u+v)\cdot\nabla\right) f(x,t) \tag{54}$$

This is anti-parabolic, so one must solve it as a final value problem which is why we didn't work with it. But that's really not a difficulty, and solving it as a final value problem, we construct a backwards Markovian propagator

$$Q_{s,t} : \mathcal{H}_t \to \mathcal{H}_s \tag{55}$$

for this equation. ($Q_{s,t}$ maps the "later" space into the "earlier" space.) It is easy to see that $(P_{t,s} f, g)_s = (f, Q_{st}\, g)_s$ for all f in $\mathcal{H}_s$ and all g in $\mathcal{H}_t$. We also obtain the direct analog of (47):

For all h in $C_b^2(\mathbb{R}^n)$ and all k in $\mathcal{H}_t$

$$(Q_{s,t} k, h)_t - (k, h)_s = \int_s^t \left(Q_{\sigma,t} k, \left(\tfrac{1}{2}\Delta - b_*\cdot\nabla\right) h\right)_\sigma d\sigma \tag{56}$$

Since $(\frac{1}{2}\Delta + b(x,t)\cdot\nabla) g(x)$ is in $\mathcal{H}_t$, we can apply this to obtain:

$$(P_{\tau,s} f, (\tfrac{1}{2}\Delta + b\cdot\nabla) g)_\tau = \int_s^t \left(Q_{\sigma\tau} (\tfrac{1}{2}\Delta + b\cdot\nabla) g, (\tfrac{1}{2}\Delta - b\cdot\nabla f)\right)_\sigma d\sigma \tag{57}$$

Combining results we have:

$$(P_{t,s} f, g)_t \leq \frac{\text{Const}}{\varepsilon^4} \int_s^t \int_s^\tau \left(1 + \int (u^2 + v^2) \rho(x,\sigma)\, dx\right) d\sigma\, d\tau \tag{58}$$

and by the finite action condition the right hand side is o(t-s) uniformly in $s \in [S,T]$. Now clearly there is a constant N so that we may cover K using no more than $N\varepsilon^{-n}$ open balls of radius $\varepsilon/2$. Crudely summing the above estimat for each of these balls we get:

$$\Pr\{\xi(s) \in K, |\xi(t) - \xi(s)| > \varepsilon\} \leq (\text{Const})(\varepsilon^{-(n+4)}) \cdot o(t-s) \tag{59}$$

This proves the lemma.

Now we are almost done. Assume K is a cube centered about the origin with edges parallel to the axes and of length L. Let $J:\mathbb{R} \to \mathbb{R}$ be a monotonically increasing smooth function so that $J(x) = x$ for $|x| < 1/2$ L and so that J is constant for $|x| > 3/4$ L and with $|J(x-y)| \leq |x-y|$ everywhere. Let X denote the $\mathbb{R}^n$ valued process with $X^i(t) = J(\xi^i(t))$. Then

$$E|X(t)-X(s)|^m = m\int_0^L \varepsilon^{m-1} \Pr\{|X(t)-X(s)|>\varepsilon\}\,d\varepsilon$$

$$\leq m\left(C\int_0^L \frac{\varepsilon^{m-1}}{\varepsilon^{n+4}}\,d\varepsilon\right)\int_\Delta^t\int_\Delta^\tau \left(1+\int(u^2+v^2)\rho(x,\sigma)\,dx\right)d\sigma\,d\tau \tag{60}$$

Taking m = n+5 so the integral can be preformed, we have $E|X(t) - X(s)|^m = o(t-s)$ uniformly in s. By a theorem of Kolmogorov, $t \mapsto X(t)$ has a continuous version.

Considering a countable nest of cubes, one sees that $t \mapsto \xi(t)$ has a continuous version. But if the canonical version of a process has a continuous version, then the canonical version is itself contionuous. Thus $\Pr(C(\mathbb{R},\dot{\mathbb{R}}^n)) = 1$. Restricting to this set, a Fubini argument we can now integrate the drift along paths, and then a Chebychev argument shows there is no probability of hitting the point at infinity. Thus $\Pr(\Omega_c) = 1$.

Henceforth let Pr denote the restriction to Ω_c.

<u>Theorem 6</u> Under Pr, the configuration process is a diffusion with forward generator $\frac{1}{2}\Delta + b\cdot\nabla$ backward generator $-\frac{1}{2}\Delta + b_*\cdot\nabla$ and density $\rho(x,t)$.

<u>Proof</u> Clearly the configuration process is Markovian under Pr by

construction. The rest follows from (47) and (56). For instance, taking $g = 1$ and f in $C_b^2(\mathbb{R}^n)$ we get

$$\int f(x)\rho(x,t)\,dx = (f,1)_t = (P_{T,t}f,1)_T = E\,f(\xi(t)) \tag{61}$$

This proves the statement about the density. Next we must show that

$$f(\xi(t)) - f(\xi(S)) - \int_S^t \left(\tfrac{1}{2}\Delta + b(\xi(\tau),\tau)\cdot\nabla\right) f(\xi(\tau))\,d\tau \tag{62}$$

is a $\mathcal{P}_t$ martingale.(The integral is taken pathwise; path continuity, the finite action condition and the Fubini theorem allow this.) Since the process is Markovian, it suffices to show that for any g in $\mathcal{H}_\Delta$

$$E\left(\left(f(\xi(t)) - f(\xi(\Delta)) - \int_\Delta^t (\tfrac{1}{2}\Delta + b\cdot\nabla) f(\xi(\tau))\,d\tau\right) g(\xi(\Delta))\right) = 0 \tag{63}$$

By the definition of the measure the left hand side is just

$$(P_{t,\Delta}g, f) - (g,f)_\Delta - \int_\Delta^t (P_{\tau,\Delta}g, (\tfrac{1}{2}\Delta + b\cdot\nabla)f)_\tau\,d\tau \tag{64}$$

which we have proved to be zero. A similar analysis takes care of the backward generator.

Having established existence, we turn to uniqueness. Let Pr^i denote the solution to the matingale problem with the approximat drift b_*^i and the density $\rho(x,T)$ at time T. Because G_{ap} is dense in $\mathcal{C}(\Omega)$, it is easy to establish that

$$P_r^i\big|_{\mathcal{P}_T} \to P_r\big|_{\mathcal{P}_T} \tag{65}$$

in the $\sigma(\mathcal{M}(\Omega), \mathcal{C}(\Omega))$ topology. Since each of these measures satisfies $Pr^i(\Omega_c) = Pr(\Omega_c) = 1$, it is then not hard to show that (65) holds also in the $\sigma(\mathcal{M}(\Omega_c), \mathcal{C}(\Omega_c))$ topology; that is, that $Pr^i \to Pr$

weakly on $\mathcal{P}_T$ in the usual sense. Clearly, the map $(v,\rho) \mapsto Pr$ which we have constructed is the unique such map with this continuity property. Francesco Guerra has found a very nice way of expressing this continuity which he has reported on at this conference [5].

Before returning to stochastic mechanics, I would like to make two remarks. First, Wei-an Zheng has recently proven an existence theorem for diffusions with singular drift [6]. His result requires more regularity of the coefficients than ours, but his approach yeilds some more information where it works: He has a beautiful result showing that under his conditions, with probability one, the paths never visit places where the density vanishes. Second, the approach presented here, unlike that in my paper [3], can easily be adapted to treat diffusions in a Riemannian manifold for rather general metrics. These results will appear elsewhere.

Now I began by talking about stochastic mechanics, and it is time to return to that subject. Suppose we have a potential V and a solution of the Schroedinger equation for it. We define u, v, and ρ in terms of ψ, and then we want to know when the conditions (30) and (31) are satisfied. A happy answer to this is that the conditions are satisfied whenever ψ_0 has a square integrable distributional gradient and V is, say, a form small potential. This just means that

$$\left|\int |\psi|^2 V(x)\, dx\right| < a \int |\nabla\psi|^2 dx + b\int |\psi|^2 dx \quad \forall\, \psi \in H^1 \tag{66}$$

for some fixed a < 1 and b > 0. Under these condition $t \mapsto \int |\nabla\psi(x,t)|^2\, dx$ is continuous and hence bounded on all finite intervals. By direct computation using the definitions

$$\int |\nabla\psi(x,t)|^2 dx = \int (u^2(x,t) + v^2(x,t))\,\rho(x,t)\, dx \tag{67}$$

and so the finite action condition is satisfied. Similarly, all the other conditions are verified in this case. Form small potentials include most potentials of physical interest, for instance Coulomb potentials. Moreover, it is imortant to remark that that we have not used the fact that v as determined by (4), (6) and (29) is a gradient. In the presence of a magnetic field, (4) has to be modified, and then v will no longer be a gradient. Our construction nonetheless applies. With this, we conclude our discussion of the existence of the diffusions of stochastic mechanics and turn to their sample path properties.

POTENTIAL SCATTERING IN STOCHASTIC MECHANICS

We now briefly turn to the second question raised in the introduction: "Do these diffusions behave in a physically reasonable manner pathwise?"

One should first remark that since it is impossiible to make any statement about paths of particles in ordinary quantum mechanics, questions of consistency between the two descriptions do not arise in this direction. Nonetheless, what the diffusions look like pathwise is of obvious importance within stochastic mechanics. We examine this question within the context of potential scattering.

It is evident that the mathematical structures of stochastic mechanics and ordinary quantum mechanics are sufficiently different that the usual objects of interest - the wave operators and the S-matrix - cannot be the focus of attention here. To orient ourselves in this new context, we begin with the simplest intuitive picture of a scattering experiment.

Suppose one has two particles, a target and a projectile. The projectile is given initial momentum p_i, so that at the beginning of the experiment it moves nearly freely toward the target from far away. After a while, the projectile enteres the region where there is significant interaction with the target. To focus on essentials, we suppose the target to be sufficiently heavy that it may be regarded as fixed, and we assume it exerts its influence on the projectile by means of a potential V(x). Eventually, the particle emerges from the region where the scattering forces are appreciable; and it settles down to free motion with a new momentum , the final momentum p_f .

We have two goals. First, to provide a precise pathwise definition of the final momentum, preferably one that is experimentally accessible. Second, to give a method for computing this final momentum given the inital state of the projectile and the interaction with the target.

The final momentum is not measured directly in scattering experiments; one only measures configurations and times directly. One method of measuring a final momentum is this: Suppose that the projectile was close to the scattering center at time 0, and suppose it is detected in a counter at $\xi \in \mathbb{R}^3$ at time T. If the distance between the target and the counter is much greater than the range of interaction, we assume that during most of its flight, the particle traveled nearly freely with momentum close to p_f. Then we should have

$$p_f \approx \frac{1}{T}\xi \tag{68}$$

Note that we are still using the convention m = 1.

Therefore, instead of studying the time evolution of the momentum itself, we study the time evolution of $1/t\,\xi(t)$. Given a potential V(x), what we want to do then is to identify those diffusions premitted under V(x) which eventually leave the region where forces are strong, and for these diffusions show that

$$\lim_{t\to\infty} \frac{1}{t}\xi(t,\omega) \equiv p_f(\omega) \tag{69}$$

exists pathwise with probability one. Shucker [7] has proved such a result in the case where the potential is identically zero. In [8] I have proved the following result.

<u>Theorem 7</u> Let V be a potential satisfying

$$(1+|x|^2)^{\beta/2}\,|V(x)| < Const, \quad (1+|x|^2)^{\beta/2}\,|\nabla V(x)| < const \tag{70}$$

for some $\beta > 3$. Furthermore, suppose there is no zero eigenvalue or zero resonance of $-\quad + V$

Let $t \mapsto \xi(t)$ be a finite action diffusion in $\mathbb{R}^3$ which is critical for the Guerra-Morato variational principle with this potential such that

$$\lim_{T\to\infty} \frac{1}{2T} \int_{-T}^{T} \Pr\{|\xi(t)| \le R\}\,dt = 0 \quad \forall R>0, \quad E|\xi(0)|^6 < \infty \tag{71}$$

Then:

$$\lim_{t\to\infty} \frac{1}{t}\,\xi(t,\omega) \equiv p_f(\omega) \quad \text{exists a.s.} \tag{72}$$

and the random variable p_f is square integrable and has the same distribution as does the quantum mechanical final momentum for the corresponding solution $t \mapsto \psi(t)$ of the Schroedinger equation.

Our method of proof is quite different from the method Shucker used to treat the case where V = 0. We will not have time to even sketch the proof here. For a fuller discussion and other refences to the literature, see [8]. The theorem stated above is the consequence of a number of lemmas, and it is not the most general result that can be obtained from them. I wish to emphasize, however, that the wave function doesn't enter the statement of the theorem until the final statement relating the stochastic mechanical final momentum to the ordinary quantum mechanical final momentum - where of course it must enter. Unfortunately, this is not true of the proof which relies heavily on an analysis of the corresponding solution of the Schroedinger equation.

This brings us to the final point. Can one study the sample path properties of stochastic mechanical diffusions by direct probabilistic methods without ever solving the Schroedinger equation to obtain the drift field of the process? One might hope to do this in the following way. Nelson has defined forward and backward stochastic time derivatives D and D_* in [ ]. There he showed that the diffusions of stochastic mechanics satisfy a stochastic analog of Newton's second equation

$$\frac{1}{2}(D_* D + D D_*)\xi(t) = -\nabla V(\xi(t)) \tag{73}$$

The left hand side is called the stochastic acceleration of $\xi(t)$. (Indeed, (73) was the original formulation of the dynamical law of stochastic mechanics.)

Our results on scattering indicate that the stochastic acceleration retains at least some of its deterministic meaning. It would be very interesting to give a proof of our last theorem based on a study of the stochastic acceleration instead of the Schroedinger equation. At present, there are no theorems at all on what the stochastic acceleration says about sample path behavior. This seems to be an interesting question in its own right. There are many problems in ordinary quantum mechanics - asymptotic completeness for more than three particles, say - which are beset with difficulties which seem to be only technical, but technically insurmountable all the same. Even if one does not accept stochastic mechanics as reasonable physics, one cannot dismiss the possibility that stochastic mechanics may provide radically new methods for studying quantum mechanical problems.

BIBLIOGRAPHY

[1] Nelson, E.: "Quantum Fluctuations", Princeton, Princeton University

Press, 1984.

[2] Guerra, F, Morato, L.: Quantization of Dynamical Systems and Stochastic Control Theory, Phys. Rev. D., 27, 1983, 1771-1786.

[3] Carlen, E.: Conservative Diffusions, Com. Math. Phys., 94, 1984, 293-315.

[4] Nelson, E.: Regular Probability Measures in Function Space, Ann. Math., Ser. 2, 69, 1959, 630-644

[5] Guerra, F.: Lecture in these procedings

[6] Zheng, W. A.: Tightness Result for Laws of Semimartingales; Application to Stochastic Mechanics., Strasbourg preprint.

[7] Shucker, D.: Stochastic Mechanics of Systems with Zero Potential, J. Func. Analysis, 38, 1980, 146-155.

[8] Carlen, E.: Potential Scattering in Stochastic Mechanics, to appear in Ann. de l'I.H.P.

[9] Nelson, E.:"Dynamical Theories of Brownian Motion", Princeton, Princeton University Press, 1967

CHARACTERISTIC EXPONENTS FOR

STOCHASTIC FLOWS

A.P. Carverhill, M.J. Chappell and K.D. Elworthy

Mathematics Institute, University of Warwick,
COVENTRY CV4 7AL, ENGLAND.

0. INTRODUCTION

A. Let X^i, $i = 1$ to m, and A denote C^∞ vector fields on a compact n-dimensional Riemannian manifold M. Consider the Stratonovich stochastic differential equation

$$dx_t = \sum_i X^i(x_t) \circ dB_t^i + A(x_t)dt \tag{1}$$

which we shall also write as

$$dx_t = X(x_t) \circ dB_t + A(x_t)dt.$$

Here $B_t = (B_t^1,\ldots,B_t^m)$ is a Brownian motion on $\mathbb{R}^m$ and $X(x):\mathbb{R}^m \to T_xM$, $x \in M$, is defined by $X(x)e_i = X^i(x)$ for $e_1,\ldots,e_m$ the standard basis of $\mathbb{R}^m$.

Let $\{F_t:t \geq 0\}$ be the solution flow of (1) i.e. if $(\Omega, \mathcal{F},\mathbb{P})$ denotes the probability space of $\{B_t:t \geq 0\}$ then

$$F_t: \Omega \times M \to M$$

and

(i) For all $x_o \in M$ and $\omega \in \Omega$, $\{F_t(\omega,x_o):t \geq 0\}$ is a solution of (1) with initial point x_o.

(ii) For all $\omega \in \Omega$

$$F_t(\omega,-):M \to M$$

is a C^∞ diffeomorphism of M onto M, continuous in t in the C^∞ topology.

We can differentiate $F_t(\omega,-)$ to get its derivative flow on the tangent bundle TM to M

$$TF_t(\omega) : TM \to TM.$$

In particular for each x in M we have the derivative as a linear map

$$T_xF_t(\omega) : T_xM \to T_{F_t(\omega,x)}M$$

of tangent spaces. Then for $v_o \in T_{x_o}M$ the TM-valued derivative process $\{v_t : t \geq 0\}$ where

$$v_t(\omega) = T_{x_o}F_t(\omega)v_o$$

satisfies a certain S.D.E. on TM

$$dv_t = \delta X(v_t) \circ dB_t + \delta A(v_t)dt \tag{2}$$

see [9], [4].

The solutions of (1) and of (2) form Markov processes. Let A and δA denote their respective infinitesimal generators , and P_t and δP_t their semigroups, $t \geq 0$.

B. Our discussion is based on the following result of A. Carverhill [2], [3]: let ρ be an invariant measure for $\{P_t : t \geq 0\}$. *Then for* (x,ω) *in a subset* Γ *of* $M \times \Omega$ *with full* $\rho \otimes \mathbb{P}$*-measure there exists a filtration of* T_xM *by linear subspaces*

$$0 = V^{r+1}_{(x,\omega)} \subset \ldots \subset V^2_{(x,\omega)} \subset V^1_{(x,\omega)} = T_xM$$

together with real numbers

$$\lambda^r_x < \ldots < \lambda^1_x$$

such that for each $j = 1$ *to* r

$$v \in V^j_{(x,\omega)} - V^{j+1}_{(x,\omega)} \iff \lim_{t\to\infty} \frac{1}{t} \log|T_xF_t(\omega)(v)| = \lambda^j_x.$$

The numbers $\{\lambda_x^j : j = 1,\ldots,r\}$ are the *characteristic*, or *Lyapunov*, *exponents* of the system (1), with respect to ρ. When A is elliptic there is a unique invariant probability measure ρ, (and it is given by a smooth density). Under these circumstances, or more generally when ρ is ergodic with respect to $\{P_t : t > 0\}$, the exponents are ρ-almost surely independent of x. For simplicity we will usually assume this is so and write λ^j for λ_x^j.

Since M is assumed compact the choice of Riemannian metric on M is irrelevant both for the norm, $| \;\; |$, above and the metric d on M used below:

If $\lambda^j < 0$ *then, for* $(x,\omega) \in \Gamma$, *the subset (stable manifold)*

$$V^j_{(x,\omega)} = \{y \in M : \overline{\lim_{t\to\infty}} \frac{1}{t} \log d(F_t(\omega,x), F_t(\omega,y)) \le \lambda^j\}$$

is an immersed manifold in M, *tangent to* $V^j_{(x,\omega)}$ *at* x.

This is a version for stochastic dynamical systems of Ruelle's stable manifold theorem for ordinary dynamical systems [19]. The programme of extending Ruelle's ergodic theory of diffeomorphisms to the stochastic case was suggested by L. Arnold at Les Houches, June 1980.

C. Figure 1 shows a computer simulation of the flow of the S.D.E. on the circle S^1 given in terms of angular coordinates θ by

$$d\theta_t = dB_t^1 - 2 \sin \theta_t dt.$$

It illustrates one sample flow $\{F_t(\omega,-) : t \ge 0\}$ evaluated at 10 different initial points $\theta_o^1,\ldots,\theta_o^{10}$. The trajectories $\{F_t(\omega,\theta_o^i) : t \ge 0\}$ are the radial projections on the circle of the curves shown in the figure (time being drawn radially outward from the circle). Being 1-dimensional there is only one exponent, and in this case it is negative. The stable manifold of the point (1,0) appears to be all of S^1

Fig. 1 A sample flow, shown with 10 initial points, of the equation on S^1 given in angular coordinates by $d\theta_t = dB_t - 2(\sin \theta_t)dt$. Time is drawn radially outwards.
{*Computer simulation by P. Townsend and D. Williams*}.

save a point near (-1,0). After some time the 10 trajectories are indistinguishable from each other. Of course, in order to see this, it is not really necessary to take more than two different initial points because of the diffeomorphism property of the flow. This example, rotated through 90°, the 'noisy North-South flow' was discussed in detail in [2], [4].

Figure 2 shows a similar simulation for the 'gradient Brownian system'

$$d\theta_t = \cos\theta_t \, dB_t^1 + \sin\theta_t \, dB_t^2$$

for a 2-dimensional Brownian motion $B = (B^1, B^2)$. Here each solution $\{\theta_t^i : t \geq 0\}$ from θ_o^i is a Brownian motion on S^1. As shown below the Lyapunov exponent is $-\frac{1}{2}$. Both simulations were prepared by P. Townsend and D. Williams at University College of Swansea.

D. In this article we shall consider the exponents of systems of mainly 3 different types: stochastic mechanical flows, gradient Brownian flows, canonical Brownian flows. The first furnishes simple examples in 1-dimension, in the second type there is a relationship between the mean exponent $\frac{1}{n}\lambda_\Sigma$ given by

$$\lambda_\Sigma = \sum_{j=1}^{r} \dim (V^j_{(x,\omega)}/V^{j+1}_{(x,\omega)}).\lambda^j$$

and the leading eigenvalue of the Laplace-Beltrami operator Δ of M, while the third type is of differential-geometric interest, and is closely analogous to the geodesic flow of Riemannian manifolds. In certain cases its stable manifold structure can be analysed completely. The first two examples are discussed again from the point of view of large deviation theory in an appendix by Elworthy and Stroock.

Fig. 2: A sample flow, shown with 10 distinct initial points, of the equation $d\theta_t = (\cos\theta_t)dB_t^1 + (\sin\theta_t)dB_t^2$. The solution from each point is a Brownian motion on S^1, (with angle as coordinate, time drawn radially outwards).

{*Computer simulation by P. Townsend and D. Williams*}.

1. STOCHASTIC MECHANICAL FLOWS IN $\mathbb{R}$.

A. The higher dimensional case and the case of stochastic mechanics on a manifold is considered again in §3C below. Here we take $M = \mathbb{R}$, overlooking its lack of compactness.

For a sufficiently regular potential $V:\mathbb{R} \to \mathbb{R}$ let E_o be the lowest eigenvalue of $(-\hbar^2/2\,\Delta + V)$ with corresponding normalized eigenfunction $\psi_o:\mathbb{R} \to \mathbb{R}$ (> 0). Thus

$$(-\frac{\hbar^2}{2}\Delta + V)\psi_o = E_o\psi_o. \tag{3}$$

The corresponding stochastic mechanical system, the "ground state process" is given by

$$dx_t = b(x_t)dt + \sqrt{\hbar}\, dB_t^1 \tag{4}$$

where $b:\mathbb{R} \to \mathbb{R}$

is given by

$$b(x) = \frac{\hbar}{2}\nabla \log \psi_o(x)^2.$$

For $x_o, v_o \in \mathbb{R}$ the derivative process $v_t = DF_t(x_o)v_o$ has equation

$$dv_t = b'(x_t)v_t\, dt$$

whence

$$v_t = v_o \exp \int_0^t b'(x_s)ds. \tag{5}$$

Since we are in dimension one there is only the top characteristic exponent λ^1, given by

$$\begin{aligned}\lambda^1 &= \lim_{t\to\infty} \frac{1}{t} \log(v_o \exp \int_0^t b'(x_s)ds) \\ &= \lim_{t\to\infty} \frac{1}{t}\int_0^t b'(x_s)ds \\ &= \int_{\mathbb{R}} b'(x)\, \rho(dx) \qquad \text{a.s.}\end{aligned} \tag{6}$$

by the ergodic theorem, or law of large numbers. Now the invariant measure ρ is given by

$$\rho(dx) = |\psi_0(x)|^2 \, dx$$

and so integrating by parts

$$\lambda^1 = -\frac{\hbar}{2}\int_{-\infty}^{\infty} (\nabla \log|\psi_0(x)|^2)\nabla|\psi_0(x)|^2 \, dx$$

$$= -2\hbar \int_{-\infty}^{\infty} \psi_0'(x)^2 \, dx. \tag{7}$$

Thus *the process is stable* i.e. $\lambda^1 < 0$, *and up to a constant factor* λ^1 *is the negative of the mean kinetic energy of the quantum mechanical particle*

$$\lambda^1 = -\frac{4}{\hbar}\left(E_0 - \int_{\mathbb{R}} V(x)\, \psi_0(x)^2 \, dx\right) \tag{8}$$

B. However was there any real reasons why we chose equation (4) rather than the equation

$$dx_t = b(x_t)dt - \sqrt{\hbar}\sin x_t \, dB_t^1 + \sqrt{\hbar}\cos x_t \, dB_t^2 \qquad (9)?$$

Both determine Markov processes with the same generator

$$A = \frac{1}{2}\hbar\Delta + b.\nabla .$$

For this equation

$$dv_t = (b'(x_t)dt - \sqrt{\hbar}\cos x_t \, dB_t^1 - \sqrt{\hbar}\sin x_t \, dB_t^2)v_t$$

giving

$$v_t = v_0 \exp\left\{\int_0^t b'(x_s)ds - \sqrt{\hbar}\int_0^t \cos x_s \, dB_s^1 - \sqrt{\hbar}\int_0^t \sin x_s \, dB_s^2 - \frac{1}{2}t\,\hbar\right\}$$

Thus the new exponent $\tilde{\lambda}^1$, say, is given by

$$\tilde{\lambda}^1 = \lim_{t\to\infty} \{\frac{1}{t}\int_0^t b'(x_s)ds - \frac{1}{t}M_t - \frac{1}{2}\hbar\}$$

where

$$M_t = \int_0^t \cos x_s \, dB_s^1 + \int_0^t \sin x_s \, dB_s^2$$

is a Brownian motion, and so has $\lim_{t\to\infty} \frac{1}{t} M_t = 0$. Arguing as before we see

$$\tilde{\lambda}^1 = \lambda^1 - \frac{1}{2}\hbar. \tag{10}$$

Since both (4) and (9) could be considered as equations on the compact manifold M this shows that *the characteristic exponents are not determined by the infinitesimal generator* A *of the process:* they depend on the S.D.E. itself. In particular if any physical significance is to be attached to them in stochastic mechanics, or anywhere else, the way in which the noise is introduced must be carefully specified. See also §3C below.

2. GRADIENT BROWNIAN FLOWS

A. A standard way of obtaining Brownian motion on M is to isometrically embed (or immerse) it in some Euclidean space $\mathbb{R}^m$ and let $X(x):\mathbb{R}^m \to T_xM$ be the orthogonal projection, or equivalently take

$$X^i = \nabla f^i$$

for $f = (f^1,\dots,f^m)$ the embedding map $f:M \to \mathbb{R}^m$. The equation

$$dx_t = \sum_{i=1}^m X(x_t)^i \circ dB_t^i$$

then has $A = \frac{1}{2}\Delta$ i.e. each solution is a Brownian motion on M, e.g. see [9] §10C or the Bibos I talk by J. Lewis.

For the case of the standard embedding of S^1 in $\mathbb{R}^2$ this equation is just (9) above with $b \equiv 0$ where x_t refers to the angle, and the argument in §1B shows that the exponent is $-\frac{1}{2}$. It is illustrated in Fig. 2. M. Chappell [7] has shown that if S^1 is isometrically embedded in $\mathbb{R}^2$ as an ellipse then the exponent of its corresponding gradient Brownian flow varies between $-\frac{1}{2}$ and $-\infty$, tending to the latter as the eccentricity of the ellipse increases.

B. For these flows, which we call *gradient Brownian flows*, we can apply Carverhill's analogue of Khasminskii's formula [4], or use Itô's formula for $\frac{1}{2}\log|v_t|^2$ directly to see that for almost all (x,ω) if $\lambda(v_0) = \lim_{t\to\infty} \frac{1}{t}\log|v_t|$, so that $\lambda(v_0) \in \{\lambda^1,\ldots,\lambda^r\}$ then

$$\lambda(v_0) = \lim_{t\to\infty} \frac{1}{t} \Big\{\frac{1}{2}\int_0^t |\alpha_{x_s}(\eta_s,-)|^2\, ds - \int_0^t |\alpha_{x_s}(\eta_s,\eta_s)|^2 ds - \frac{1}{2}\int_0^t \mathrm{Ric}(\eta_s,\eta_s)ds\Big\} \text{ a.s.} \tag{11}$$

where Ric $(-,-)$ refers to the Ricci tensor of M, and

$$\alpha_x : T_xM \times T_xM \to T_xM^{\perp} \subset \mathbb{R}^m \qquad x \in M$$

is the second fundamental form [12], [9], and η_s is the tangent sphere bundle valued process

$$\eta_s = v_s/|v_s|.$$

Recall that

$$\mathrm{Ric}\,(v,v) = -|\alpha_x(v,-)|^2 + \langle\alpha_x(v,v),\, nN_x\rangle,\ v \in T_xM \tag{12}$$

where

$$N_x = \frac{1}{n}\,\mathrm{trace}\,\alpha_x \in T_xM^{\perp}$$

is the mean curvature normal.

To obtain precise information from (11) and its analogues for other S.D.E's one needs to know about the invariant measures for the process $\{\eta_s : s \geq 0\}$, [5], [11]. This does not seem easy when $n > 1$.

For the sphere $S^n(r)$ of radius $r > 0$ with its standard embedding in $\mathbb{R}^{n+1}$ we have, for $u, v \in T_x S^n(r)$,

$$\mathrm{Ric}\,(u,v) = \frac{n-1}{r^2} \langle u,v\rangle, \quad \alpha_x(u,v) = -\frac{\langle u,v\rangle}{r} \cdot \frac{x}{r}$$

whence

$$\lambda^1 = -\frac{1}{2}\frac{n}{r^2}.$$

Thus *the gradient Brownian flow on* $S^n(r)$ *is stable with the maximum exponent equal to the leading eigenvalue of one half the Laplace-Beltrami operator of* $S^n(r)$. In fact there are no other exponents because the mean exponent $\frac{1}{n}\lambda_\Sigma$ is also the leading eigenvalue: see §3 below.

3. MEAN EXPONENTS

A. Baxendale pointed out to us that there is a formula for the weighted sum of the exponents λ_Σ, (see §0.C), which needs only knowledge of the invariant measure ρ of A and not those of any derivative systems. Using this together with results of Reilly [18], Chappell [8] was able to show that *for any compact* M *embedded in* $\mathbb{R}^m$ *the induced gradient Brownian flow satisfies:*

$$\textit{the mean exponent } \frac{1}{n}\lambda_\Sigma \le \textit{the leading eigenvalue of } \frac{1}{2}\Delta \qquad (13)$$

B. From the proof of the multiplicative ergodic theorem, [19] and its extensions [2], [3] to our situation we have

$$\lambda_\Sigma = \lim_{t\to\infty} \frac{1}{t} \log |\det T_x F_t(\omega)|$$

for $\rho \otimes \mathbb{P}$-almost all (x,ω), where the determinant is taken using the Riemannian inner products on the tangent spaces at x and $F_t(\omega,x)$. Thus λ_Σ has the geometric interpretation as the exponential rate at which the flow changes volume (or area if dim M = 2).

Itô formulae for the determinant have been obtained by various authors, in particular Malliavin [16]. An easy way is to observe that the S.D.E. (2) for the derivative flow determines an S.D.E. on the manifold

$$\mathbb{L}(TM;TM) = \bigcup_{x,y\in M} \mathbb{L}(T_xM;T_yM)$$

consisting of the disjoint union of the spaces of linear maps of T_xM into T_yM. The precise form of this S.D.E. is irrelevant. We then have the smooth function

$$|\det| : \mathbb{L}\text{is}\ (TM;TM) \to \mathbb{R}$$

on the open subset of $\mathbb{L}(TM;TM)$ consisting of invertible elements. If we apply the form of the Itô formula given in Lemma VII 9B(ii) of [9], use of the classical continuity equation for the Jacobian determinants of flows of ordinary dynamical systems quickly yields

$$\log |\det T_{x_0}F_t| = \int_0^t \sum_i \operatorname{div} X^i(x_s)dB_s^i + \int_0^t \operatorname{div} A(x_s)ds + \frac{1}{2}\int_0^t \sum_i \langle \nabla \operatorname{div} X^i(x_s), X^i(x_s)\rangle ds \qquad (14)$$

which gives Baxendale's formula

$$\lambda_\Sigma = \int_M \operatorname{div} A(x)\rho(dx) + \frac{1}{2}\int_M \sum_i \langle \nabla \operatorname{div} X^i(x), X^i(x)\rangle\, \rho(dx) \qquad (15)$$

When $A = \frac{1}{2}\Delta$, i.e. for a flow of Brownian motions, $\rho(dx) = |M|^{-1}dx$ where $|M|$ is the volume of M and we can integrate by parts and apply the divergence theorem to obtain

$$\lambda_\Sigma = -\frac{1}{2|M|}\int_M \sum_i (\operatorname{div} X^i(x))^2 dx \qquad (15)$$

It is not clear to us whether, for an arbitrary compact Riemannian manifold M, it is possible to choose $X^1,\dots,X^m$ with $\operatorname{div} X^i = 0$ for each i, and with $A = \frac{1}{2}\Delta_M$ for a suitable choice of A. This last condition is simply that for each $x \in M$ the corresponding $X(x):\mathbb{R}^m \to T_xM$ should be surjective and induce the given inner product on T_xM. If not one can ask what the maximum value of λ_Σ can be over all such choices of $X^1,\ldots,X^m$, for m allowed to vary, and whether this number has significance in other directions.

A corollary of the Hodge decomposition theorem extends Helmholtz's theorem from the case $M = \mathbb{R}^3$ to enable us to write X^i uniquely as the sum of a divergence free field Y^i and a gradient, [13]: $X^i = Y^i + \nabla\phi^i$ say, for

$$\phi^i : M \to \mathbb{R},$$

with $$\int_M \phi^i = 0.$$

Then (15) becomes

$$\lambda_\Sigma = -\frac{1}{2|M|}\int_M \sum_i (\Delta\phi^i)^2$$

$$= -\frac{1}{2|M|}\int_M \sum_i \langle (d^* + d)^2\, d\phi^i, d\phi^i\rangle$$

where d^* is the L^2 adjoint of the exterior differentiation operator d. Now $-(d^* + d)^2$ maps the space of exact 1-forms to itself and is conjugate on that space to the restriction of Δ to the L^2 orthogonal conjugate of the space of constant functions. Therefore

$$\lambda_\Sigma \le \frac{\mu}{|M|} \sum_i \int_M |d\phi^i|^2 \tag{16}$$

where $\mu < 0$ is the leading eigenvalue of $\frac{1}{2}\Delta$, with equality if and only if

$$\frac{1}{2}\Delta\phi^i = \mu\phi^i \qquad i = 1 \text{ to } m. \tag{17}$$

For a gradient Brownian flow $X^i = \nabla f^i$. Therefore $Y^i \equiv 0$ and

$$\phi^i = f^i - \int_M f^i \frac{dx}{|M|}$$

with

$$\Sigma|d\phi^i|^2 = \Sigma|X^i|^2 = n.$$

Therefore (16) reduces to Chappell's upper bound (13). Moreover by [12] Note 14, or [20], we have Takahashi's result that (17) is true, and equivalently *equality holds in* (13), *if and only if* M *is embedded as a minimal submanifold of some hypersphere in* $\mathbb{R}^m$ *of radius* $\sqrt{(\frac{1}{2}n\,|\mu|^{-1})}$.

Chappell's estimates [8] for λ_Σ were obtained from the fact that in the gradient Brownian case (15) becomes

$$\lambda_\Sigma = \frac{-1}{2|M|} \int_M (\text{trace } \alpha_x)^2 \, dx.$$

C. For a stochastic mechanical flow in $\mathbb{R}^n$ (or the flat torus $\mathbb{T}^n$ to be safely compact) we can take the multi-dimensional versions of equations (3), (4) so that (15) reduces to the analogue of (6):

$$\begin{aligned} \lambda_\Sigma &= \int_{\mathbb{R}^n} (\text{div } b(x))\ \rho(dx) \\ &= -2\hbar \int_{\mathbb{R}^n} |\nabla\psi_0(x)|^2 \, dx \end{aligned}$$

as in (7). Thus *up to a constant factor the sum* λ_Σ *of the exponents is the negative of the mean kinetic energy of the quantum mechanical particle*

$$\lambda_\Sigma = -\frac{4}{\hbar}\left(E_0 - \int_{\mathbb{R}^n} \psi_0(x)\ V(x)\ \psi_0(x)dx\right). \tag{18}$$

(Strictly speaking we have no right, for the noncompact manifold $\mathbb{R}^n$, to assert that the sum λ_Σ is $\lim_{t\to 0} \frac{1}{t} \log |\det T_x F_t(\omega)|$ and it is the latter which we have computed.)

For a stochastic mechanical flow on an arbitrary compact Riemannian manifold M there is not quite such a natural choice of stochastic differential equation and different choices will give different values to λ_Σ as described in §1.B. However there is the canonical method of lifting everything to the total space of the frame bundle, as described in the next section. Using the results in [17] and equation (15) it is easy to see that (18) still holds for the flow on the frame bundle.

4. CANONICAL FLOWS

A. There is a canonical way to assign a stochastic flow to a diffusion generator $A = \frac{1}{2}\Delta + A$, for A a smooth vector field on our Riemannian manifold M. However the flow is a flow of diffeomorphisms of the space O(M) of orthonormal frames of M, not of M itself. We will consider a frame as an isometry, u say, of $\mathbb{R}^n$ with its standard inner product, to T_xM with its Riemannian inner product, for some $x \in M$. Thus we have a projection

$$\pi: O(M) \to M$$
$$\pi(u) = x.$$

The Levi-Civita connection on M determines linear maps

$$X(u): \mathbb{R}^n \to T_uO(M) \qquad u \in O(M)$$

such that

$$T\pi(X(u)e) = u(e) \qquad e \in \mathbb{R}^n$$

and also a 'horizontal lift' of A to a vector field $\tilde{A}$ on O(M) such that

$$T\pi(\tilde{A}(u)) = A(\pi(u)) \qquad u \in O(M)$$

As described in [9] or [10] it turns out that if we choose $u_o \in \pi^{-1}(x_o)$ for $x_o \in M$ and solve

$$du_t = X(u_t) \circ dB_t + \tilde{A}(u_t)dt \tag{19}$$

then $x_t \equiv \pi(u_t)$ is a Markov process on M, with marginal distributions independent of u_o, and with differential generator $\frac{1}{2}\Delta + A$. The flow F of (19) will be called the *canonical flow* when A = 0, and the *canonical flow with drift* A otherwise. The characteristic exponents depend only on the operator $\frac{1}{2}\Delta + A$, and so when A = 0 only on the Riemannian manifold itself. It is shown in [6] that they can be taken to be independent of u in O(M).

B. At each point u of $\pi^{-1}(x)$, for $x \in M$, there is a decomposition of $T_uO(M)$ into horizontal vectors and vertical vectors. The former are the image of X(u) and the latter tangent vectors to $\pi^{-1}(x)$, which is diffeomorphic to the orthogonal group O(n). Using this decomposition we can represent TO(M) as

$$TO(M) \cong O(M) \times \mathbb{R}^n \times \underline{o}(n) \tag{20}$$

where $\underline{o}(n)$ is the Lie algebra of O(n), and will be identified with the space of skew-symmetric $n \times n$-matrices. The Hilbert-Schmidt inner product on $\underline{o}(n)$ and the Euclidean inner product on $\mathbb{R}^n$ give a Riemannian structure to O(M) via (20). The corresponding volume element on O(M) is known as the Liouville volume element. If dim M = 2 each connected component of O(M) can be identified with the unit sphere bundle of M, for connected M, and this measure corresponds to the usual Liouville measure.

With respect to this metric each X^i is divergence free, and $\tilde{A}$ is if A was. See [17]. Thus the canonical flow is measure preserving almost surely, as observed by Malliavin [16]. Thus *for the canonical flow* $\lambda_\Sigma = 0$. (21)

C. For $v_o \in T_{u_o}O(M)$ we can use (20) to represent $v_t \equiv TF_t(v_o)$ as (ξ_t, A_t) with values in $\mathbb{R}^n \times \underline{o}(n)$. Coupled stochastic integral equations for ξ_t and A_t are given in [9] Chapter VII §12. It is shown in [6] that ξ_t almost surely never vanishes if $\xi_0 \neq 0$ almost surely, and the equations yield, for dim M = 2,

$$\log |\xi_t| = \log |\xi_0| + \int_0^t \left\langle \frac{\xi_s}{|\xi_s|}, \frac{A_s}{|\xi_s|} dB_s \right\rangle - \frac{1}{2} \int_0^t S(x_s) ds \tag{22}$$

where $S: M \to \mathbb{R}$ is the Gaussian curvature of M.

Let $\{M_t : t \geq 0\}$ denote the stochastic integral in (22). Since it is a time-changed Brownian motion

$$\overline{\lim_{t\to\infty}} \frac{1}{t} M_t \geq 0 \quad \text{a.s.} \tag{23}$$

$$\text{Thus } \lambda^1 \geq \overline{\lim_{t\to\infty}} \frac{1}{t} \log |\xi_t| \geq -\frac{1}{2} \overline{\lim_{t\to\infty}} \frac{1}{t} \int_0^t S(x_s) ds$$

$$= -\frac{1}{2} \int_M S(x) \frac{dx}{|M|} \tag{24}$$

by the ergodic theorem. Applying the Gauss-Bonnet theorem we obtain one of the results of [6]: *If* dim M = 2 *then*

$$\lambda^1 \geq -\frac{\pi}{|M|} \chi(M) \tag{25}$$

where $\chi(M)$ *is the Euler characteristic of* M. For constant negative curvature there is equality in (24) and (25). There is an analogous formula to (24) in higher dimensions in terms of the Ricci curvature [6].

D. For manifolds of constant curvature it is possible to get precise and detailed results about all the exponents and the stable manifolds, [1], [6]. For this it is easiest to replace M by its simply connected covering space $\tilde{M}$ which will be either a sphere S^n or a hyperbolic space H^n. These are diffeomorphic to $SO(n+1)/SO(n)$ and $O_+(1,n)/SO(n)$ respectively where $O_+(1,n)$ is the identity component of the Lorentz group in 1 time and n space dimensions. In fact the quotient maps

$$SO(n+1) \to SO(n+1)/SO(n),$$

and

$$O_+(1,n) \to O_+(1,n)/SO(n)$$

can be identified with the projection map of the special orthogonal frame bundle $SO(\tilde{M}) \to \tilde{M}$ [12], page 268, and the canonical flow on $SO(\tilde{M})$ is given by

$$du = \tilde{X}(u) \circ dB \qquad (26)$$

where $\tilde{X}$ is left invariant i.e. each $\tilde{X}^i$ is a left invariant vector field. The flow $\tilde{F}_t(\omega): SO(\tilde{M}) \to SO(\tilde{M})$ is therefore given in terms of the group multiplication by

$$\tilde{F}_t(\omega)(u) = u.g_t(\omega)$$

where $\{g_t : t \geq 0\}$ is the solution of (26) with g_o the identity in the group.

Now SO(n+1) has a bi-invariant Riemannian metric, and so with respect to that each $\tilde{F}_t(\omega)$ consists of isometries. Thus *for constant positive curvature there is only the exponent* $\lambda^1 = 0$. A fact that can also be easily seen from the formulae for ξ_t and A_t mentioned above, [6].

The case of hyperbolic space H^n is more interesting. It turns out that there are 3 exponents $\lambda^1 = \frac{n-1}{2}$, $\lambda^2 = 0$, $\lambda^3 = -\frac{(n-1)}{2}$ and that the stable manifold corresponding to λ^3 is, just as for the geodesic flow, obtained by looking at the point at infinity to which $\pi F_t(\omega, u_o)$ goes as $t \to \infty$, and taking the horocycle subgroup corresponding to that point. These results were suggested to us by Guivar'ch after his consideration of the discrete time case; the n = 2 case is treated by a different method in [6], while the most complete results for general n are obtained by these methods in [1]. See also [14], [15] where long time behaviour is discussed for more general symmetric spaces.

ACKNOWLEDGEMENTS

Suggestions from L. Arnold started our interest in this project. Discussions with G. Jona-Lasinio about stochastic mechanics and with I. Guivar'ch about canonical flows proved very helpful, as did discussions with P. Baxendale. We would like to thank D. Williams and P. Townsend for permission to reproduce their computer simulations. The last named author wishes to thank IRMA Rennes for their hospitality in September 1983. This research was partially supported by SERC grants GR/C/13644 and GR/C/60860. The typing was by Terri Moss.

REFERENCES

[1] Baxendale, P.H. (1984). Asymptotic behaviour of stochastic flows of diffeomorphisms: two case studies. Preprint: Dept. of Maths., Univeristy of Aberdeen, Scotland.

[2] Carverhill, A.P. (1983). Flows of stochastic dynamical systems: ergodic theory of stochastic flows. Ph.D. Thesis, University of Warwick, Coventry, England.

[3] Carverhill, A.P. (1983). Flows of stochastic dynamical systems: Ergodic Theory. To appear in Stochastics.

[4] Carverhill, A.P. (1984). A formula for the Lyapunov numbers of a stochastic flow. Application to a perturbation theorem. To appear in Stochastics.

[5] Carverhill, A.P. (1984). A "Markovian" approach to the multiplicative ergodic theorem for nonlinear stochastic dynamical systems. Preprint: Mathematics Institute, University of Warwick,

Coventry CV4 7AL, England.

[6] Carverhill, A.P. and Elworthy, K.D. (1985). Lyapunov exponents for a stochastic analogue of the geodesic flow. Preprint: Mathematics Institute, University of Warwick, Coventry CV4 7AL.

[7] Chappell, M.J. (1984). Lyapunov exponents for gradient Brownian systems on the circle. Research Report, Mathematics Department, University of Warwick.

[8] Chappell, M. (1984). Bounds for average Lyapunov exponents of gradient stochastic systems. To appear in proceedings of 'Workshop on Lyapunov exponents' Bremen. November 1984. Lecture Notes in Mathematics. Springer-Verlag.

[9] Elworthy, K.D. (1982). "Stochastic Differential Equations on Manifolds. London Math. Soc." Lecture Notes in Mathematics. Cambridge University Press.

[10] Ikeda, N. and Watanabe, S. (1981). Stochastic Differential Equations and Diffusion Processes. Tokyo: Kodansha. Amsterdam, New York, Oxford: North-Holland.

[11] Kifer, Yu. (1984). A Multiplicative Erogdic Theorem for Random Transformations. Preprints: Institute of Mathematics. Hebrew University of Jerusalem, Jerusalem.

[12] Kobayashi, S. and Nomizu, K. (1969), Foundations of differential geometry, Vol. II. New York, Chichester, Brisbane, Toronto: Interscience Publishers, John Wiley & Sons.

[13] Marsden, J.E, Ebin, G.E., and Fischer A.E. (1972). Diffeomorphism groups, hydrodynamics and relativity. In Proc. 13th Biennial Seminar of Canadian Math. Congress, Halifax 1971, Vol. I, ed. J.R. Vanstone, pp. 135-279. Montreal: Canadian Mathematical Congress.

[14] Malliavin M.-P. and Malliavin, P. (1974). Factorisations et lois limites de la diffusion horizontale au-dessus d'un espace Riemannien symmetrique. In Theory du Potential et Analyse Harmonique. ed. J. Faraut, Lecture Notes in Maths. 404. Springer-Verlag.

[15] Malliavin, M.-P. and Malliavin, P. (1975). Holonomic stochastique au-dessus d'un espace Riemannien symetrique. C.R. Acad. Sc. Paris, 280, Serie A, 793-795.

[16] Malliavin, P. (1977). Champs de Jacobi stochastques. C.R. Acad. Sc. Paris, 285, Serie A, 789-792.

[17] O'Neill, B. (1966). The fundamental equations of a submersion, Michigan Math. J., 13, 459-469.

[18] Reilly, R.C. (1977). On the first eigenvalue of the Laplacian for compact submanifolds of Euclidean space. Comment. Math. Helvetici, 52, 525-533.

[19] Ruelle, D. (1978), Ergodic Theory of Differentiable Dynamical Systems. Publications I.H.E.S., Bures-Sur-Yvette, France.

[20] Takahashi, T. (1966). Minimal immersions of Riemannian manifolds. J. Math. Soc. Japan, 4, 380-385.

APPENDIX : LARGE DEVIATION THEORY FOR MEAN EXPONENTS OF STOCHASTIC FLOWS

K.D. Elworthy, Mathematics Institute, University of Warwick, Coventry CV4 7AL, England.
D. Stroock, Department of Mathematics, M.I.T. Cambridge, Mass. U.S.A.

A. Let M be a connected, compact manifold of dimension n with vector fields $X^1,\dots,X^m$ and A satisfying $\mathrm{Lie}(X^1,\dots,X^m)(x) = T_xM$ for every x in M. Define $(t,x_o) \to F_t(x_o)$ as the solution flow of

$$dx_t = \sum_1^m X^i(x_t) \circ dB_t^i + A(x_t)dt.$$

Set $x_t = F_t(x_0)$ and denote by $\{P_t : t > 0\}$ the associated semi-group. The following are applications of the regularity theory and the strong maximum principle for solutions to problems involving the operator $A = \frac{1}{2}\sum_1^m (X^i)^2 + A$; the first is standard, and the second comes from [4] (see Remark 2.43 there) where there is a detailed proof.

(i) There is a unique probability measure μ_0 on M such that $\mu_0 P_t = \mu_0$, $t > 0$. Moreover $\mu_0(dx) = \lambda_0(x)dx$ where λ_0 is a positive element of $C^\infty(M)$

(ii) Given $\{\sigma^*,\ldots,\sigma^m\} \subseteq C^\infty(M)$, set

$$\gamma = \inf\{ \sum_1^m \int (X^i\phi - \sigma_i)^2 \, d\mu_0 : \phi \in C^\infty(M)\}.$$

If $\gamma = 0$, then there is a unique $f \in C^\infty(M)$ such that $\int f\, d\mu_0 = 0$ and $X^i f = \sigma_i$, $1 \le i \le m$.

B. For given elements $\sigma^1,\ldots,\sigma^m$ and Q of $C^\infty(M)$, define $(t,x_0) \to \rho(t,x_0)$ by

$$\rho(t,x_0) = \sum_1^m \int_0^t \sigma^i(x_s)dB_s^i + \int_0^t Q(x_s)ds$$

and define γ as in (ii). If $\gamma > 0$, define $I:\mathbb{R} \to [0,\infty)$ by

$$I(\rho) = \sup_\phi \inf_\mu \frac{(\rho - \int (Q-A\phi)d\mu)^2}{2\Sigma\int(\sigma^i - X^i\phi)^2 d\mu}$$

taken over μ in $M_1(M)$, the probability measures on M, and $\phi \in C^\infty(M)$. Here and below $\beta^2/0$ is interpreted as $+\infty$ if $\beta \neq 0$ and as 0 when $\beta = 0$.

If $\gamma = 0$, let f be as in (ii), set $\tilde{Q} = Q - Af$, and define

$$I:\mathbb{R} \to [0,\infty) \cup \{\infty\}$$

by

$$I(\rho) = \inf_\mu \{J(\mu) : \mu \in M_1(M) \;\&\; \rho = \int \tilde{Q}\, d\mu\}$$

where

$$J(\mu) = \sup \{ \frac{(\int A\phi \, d\mu)^2}{2\Sigma \int (X^i\phi)^2 d\mu} : \phi \in C^\infty(M) \}.$$

The following summarizes some of the principle results of [4] set into this context, as described in Remark 2.44 of [4].

<u>Theorem</u> (Stroock [4]): The function I is lower semi-continuous and convex, $I(\int Q \, d\mu_o) = 0$ and there is an $\varepsilon > 0$ such that $I(\rho) \geq \varepsilon(\rho - \int Q \, d\mu_o)^2$ for all $\rho \in \mathbb{R}$. Moreover: if $\gamma > 0$, then $I(\rho) \leq (2\gamma)^{-1} (\rho - \int Q \, d\mu_o)^2$ for all $\rho \in \mathbb{R}$, and if $\gamma = 0$ then I is continuous on $(\text{Range } \tilde{Q})^o$ and takes the value $+\infty$ off of Range $\tilde{Q}$. Finally for any Borel set Γ of $\mathbb{R}$:

$$\begin{aligned} -\inf_{\Gamma^o} I &\leq \underline{\lim}_{T\to\infty} \inf_{x_o \in M} \frac{1}{T} \log \mathbb{P}(\rho(T,x_o)/T \in \Gamma) \\ &\leq \overline{\lim}_{T\to\infty} \sup_{x_o \in M} \frac{1}{T} \log \mathbb{P}(\rho(T,x_o)/T \in \Gamma) \\ &\leq -\inf_{\bar{\Gamma}} I. \end{aligned}$$

In particular, if $\bar{\Gamma} = \overline{\Gamma^o}$ and either $\gamma > 0$ or $\partial\Gamma \cap \partial(\text{Range } \tilde{Q}) = \emptyset$, then

$$\lim_{T\to\infty} \sup_{x_o \in M} \left| \frac{1}{T} \log \mathbb{P}(\rho(T,x_o)/T \in \Gamma) + \inf_{\Gamma} I \right| = 0. \qquad //$$

Now furnish M with a Riemannian metric and associated Levi-Civita connection and set

$$\lambda_\Sigma(t,x_o) = \log|\det T_{x_o} F_t|.$$

Then, from equation (14):

<u>Corollary</u>: Set $\sigma^i = \text{div } X^i$, $1 \leq i \leq m$, and

$$Q = \text{div}\, A + \frac{1}{2} \sum_{1}^{m} \langle \nabla \text{div}\, X^i, X^i \rangle.$$

If I is defined accordingly, then the preceding applies with $\rho(t,x) = \lambda_\Sigma(t,x)$. //

Remark 1(i) When $\sum_{i=1}^{m} c^i X^i = 0$ the expression for I simplifies and becomes:

$$I(\rho) = \inf_{\mu} \left\{ \frac{(\rho - \int Q d\mu)^2}{2\int \Sigma\sigma_i^2\, d\mu} + J(\mu) : \mu \in M_1(M) \right\}$$

The proof of this comes from equation (2.14) of [4], the expression given there for $\Lambda(\lambda)$ in (2.25), the convexity of $\mu \to J(\mu)$, and the mini-max theorem.

(ii) If Q and $\sum_{1}^{m} \sigma_i^2$ are constant, with values $\bar{Q}$ and $\bar{a}$ say, then $t \to \rho(\bar{a}^{-1} t, x_o)$ is a Brownian motion with constant drift $\bar{a}^{-1}\bar{Q}$ and so

$$I(\rho) = \frac{(\rho - \bar{Q})^2}{2\bar{a}} .$$

Remark 2. When A is symmetric in $L^2(\mu_o)$, i.e. $\int \phi\, A\psi\, d\mu_o = \int \psi\, A\phi\, d\mu_o$ for ϕ, ψ in $C^\infty(M)$, let E denote the Dirichlet form associated with A:

$$E(\phi,\psi) = -\int \phi A\psi\, d\mu_o = \sum_{1}^{m} \int (X^i\phi)(X^i\psi) d\mu_o$$

for ϕ, ψ in $C^\infty(M)$. Then

$$J(\mu) = \begin{cases} E\left(\sqrt{\frac{d\mu}{d\mu_o}}, \sqrt{\frac{d\mu}{d\mu_o}}\right) & \text{if} \quad \mu << \mu_o \\ \infty \text{ otherwise.} \end{cases}$$

This result can be found in Donsker-Varadhan [2] and is Theorem (7.44), page 152 of [3].

C. Example 1 : Stochastic Mechanics on $\mathbb{T}^n$

Consider the Hamiltonian $H = -\frac{\hbar^2}{2}\Delta + V$ on the flat torus $\mathbb{T}^n$. As in §§1A, 1C take the ground stake ψ_0:

$$(-\frac{\hbar^2}{2}\Delta + V)\psi_0 = E_0\psi_0$$

and ground state stochastic differential equation

$$dx_t = A(x_t)dt + \sqrt{\hbar}\, dB_t$$

where

$$A(x) = \frac{\hbar}{2}\nabla\log\psi_0(x)^2.$$

Thus $\sigma^i \equiv 0$ for each i and $\gamma = 0$ with corresponding function $f \equiv 0$. The invariant measure is $\mu_0(dx) = \psi_0(x)^2 dx$ and A is symmetric, being conjugate to $h^{-1}(E_0 - H)$, with Dirichlet form

$$E(\phi,\phi) = \frac{h}{2}\int|\nabla\phi|^2\, d\mu_0.$$

If $\mu(dx) = \psi(x)^2 dx$ for smooth $\psi:\mathbb{T}^n \to \mathbb{R}(\geq 0)$ it follows from Remark 2, or a direct computation using the Cauchy-Schwartz inequality, that

$$J(\mu) = E(\psi,\psi)$$

$$= \frac{1}{\hbar}\int\psi(H-E_0)\psi\, dx.$$

Now let $\lambda(p)$ be the leading eigenvalue of $A(p)$ where

$$A(p) = \frac{1}{2}\hbar\Delta \;+ A. \;+ p\,\mathrm{div}\,A \qquad p \in \mathbb{R}.$$

Set $\qquad H(p) = h - p\hbar\,\mathrm{div}\,A \qquad p \in \mathbb{R}.$

The conjugacy (via $\psi \mapsto \psi\psi_0$) of $A(p)$ with $-\hbar^{-1}(H(p)-E_0)$ shows that $E_0 - \lambda(p)\hbar$ is the least eigenvalue of $H(p)$. The minimum, $\tilde{I}(\rho)$ say, of $J(\mu)$ over all $\mu \in M_1(\mathbb{T}^n)$ of the form $d\mu = \psi(x)^2 dx$ with ψ smooth and satisfying the constraint $\rho = \int(\mathrm{div}\,A)\psi^2 dx$ is attained when ψ is an eigenvector of $H(p_\rho)$ for some p_ρ. Let $E_0 - \tilde{\lambda}(\rho)h$ be the corresponding eigenvalue. Then

$$\tilde{I}(\rho) = p_\rho \rho - \tilde{\lambda}(\rho) \le p_\rho \rho - \lambda(p_\rho)$$

However it is shown in [4] that $I(\rho)$ is the Legendre transform

$$I(\rho) = \sup_p \{p\rho - \Lambda(p)\}$$

where

$$\begin{aligned}\Lambda(p) &= \lim_{t\to\infty} \frac{1}{t} \log \mathbb{E} \exp (p\, \lambda_\Sigma(t,x)) \\ &= \lim_{t\to\infty} \frac{1}{t} \log \mathbb{E} \exp\left(p \int_0^t \operatorname{div} A(x_s)\,ds\right) \\ &= \lim_{t\to\infty} \frac{1}{t} \log \mathbb{E} \,|\det T_x F_t|^p\end{aligned}$$

which is the 'p-th moment exponent', [1], for the determinant. Now as shown in [1], or as follows from the second expression above for $\Lambda(p)$ by Donsker and Varadhan's extensions of Kac's result for Brownian motion, $\Lambda(p)$ is just the leading eigenvalue $\lambda(p)$ of $\frac{1}{2}\Delta + A + p \operatorname{div} A$. Also by definition $I(\rho) \le \tilde{I}(\rho)$. Thus

$$\sup_p \{p\rho - \lambda(p)\} = I(\rho) \le \tilde{I}(\rho) \le p_\rho \rho - \lambda(p_\rho)$$

showing in particular that

$$I(\rho) = \tilde{I}(\rho).$$

D. Example 2: Gradient Brownian Flows

As in §2 consider an isometric immersion

$$f = (f^1,\ldots,f^m): M \to \mathbb{R}^m.$$

Take $X^i = \nabla f^i$ and $A = 0$. Then $\mathcal{A} = \frac{1}{2}\Delta$ and $d\mu_0 = |M|^{-1}dx$. For e in $\mathbb{R}^m$ let X^e denote the vector field $X(\cdot)^e$. When e has norm one $\operatorname{div} X^e(x)$ is just the component of the trace of the second fundamental form α_x

in the direction e, since

$$\operatorname{div} X^e = \operatorname{trace} \nabla X^e = \langle (\operatorname{trace} \nabla^2 f), e \rangle .$$

In particular it vanishes for e tangent to M at x, while $X^e(x)$ vanishes for e normal at x. Then $\Sigma(\operatorname{div} X^i)X^i \equiv 0$ and a quick calculation shows that

$$\gamma = |M|^{-1} \int |\operatorname{trace} \alpha_x|^2 dx = -2\lambda_\Sigma > 0.$$

Moreover, taking divergences

$$\Sigma(\operatorname{div} X^i)^2 = - \Sigma\langle \nabla \operatorname{div} X^i, X^i \rangle .$$

By Remark 1 this gives

$$I(\rho) = \inf_\mu \left\{ \frac{(\rho + \frac{1}{2}\int |\operatorname{trace} \alpha_x|^2 \mu(dx))^2}{2\int |\operatorname{trace} \alpha_x|^2 \mu(dx)} + J(\mu) \right\}$$

where

$$J(\mu) = \sup \left\{ \frac{1}{8} \frac{(\int \Delta\phi \, d\mu)^2}{\int |\nabla\phi|^2 d\mu} : \phi \in C^\infty(M) \right\}.$$

Consequently, taking $\mu(dx) = |M|^{-1} dx$

$$I(\rho) \leq (\rho - \lambda_\Sigma)^2/(4\lambda_\Sigma)$$

with equality, as in Remark 1, when the immersion has mean curvature normal of constant length, i.e. $|\operatorname{trace} \alpha_x| = c$ some constant c. This condition holds for the standard embedding of S^n in $\mathbb{R}^{n+1}$, and for all immersions of constant mean curvature (it is equivalent to constant mean curvature for hypersurfaces of $\mathbb{R}^3$). In particular it holds for the class of immersions described in §3B for which $\lambda_\Sigma = n\mu$ for μ the leading eigenvalue of $\frac{1}{2}\Delta$, since for these

$$\frac{1}{2} \Delta\phi^i = \mu\phi^i \qquad i = 1 \text{ to } m$$

and

$$\sum_i |\phi^i|^2 = \text{const.}$$

where

$$\phi^i = f^i - \int_M f^i \frac{dx}{|M|} .$$

Thus *for this special class of immersions*

$$I(\rho) = -(n\mu-\rho)^2/(4n\mu)$$

and, for example, taking $\rho = 0$

$$\lim_{t\to\infty} \frac{1}{t} \log \mathbb{P}\{|\det T_{x_0} F_t| > 1\} = \frac{1}{4} n\mu$$

uniformly in $x_0 \in M$.

For other immersions $\lambda_\Sigma < n\mu$ and so if f is an immersion, not of the special class, but with mean curvature of constant length, then

$$\lim_{t\to\infty} \frac{1}{t} \log \mathbb{P}\ \{|\det T_{x_0} F_t| > 1\} = -\frac{1}{8|M|} \int |\text{trace}\ \alpha_x|^2 dx$$

$$< \frac{1}{4} n\mu .$$

[1] Arnold, L. (1983). A formula connecting sample and moment stability of linear stochastic systems. Report 92, Forschungsschwerpunkt Dynamische Systeme, Universität Bremen. Also article in these proceedings.

[2] Donsker, M.D. and Varadhan, S.R.S. (1975). Asymptotic evaluation of certain Markov process expectations for large time, I. Comm. Pure Appl. Math., 28, 1-47.

[3] Stroock, D. (1984). An introduction to the theory of large deviations. Universitext Series. Springer-Verlag.

[4] Stroock, D. (1984). On the rate at which a homogeneous diffusion approaches a limit, an exercise in the theory of large deviations. To appear in Ann. of Prob.

ELECTRIC FIELD AND EFFECTIVE DIELECTRIC CONSTANT IN RANDOM MEDIA WITH NON-LINEAR RESPONSE

G.F. Dell'Antonio*)

Dipartimento di Matematica, Istituto G. Castelnuovo

Università di Roma - La Sapienza

and

International School for Advanced Studies, Trieste, Italy

Introduction

In this talk I will outline some results on the existence and the average properties of a solution of the time-independent Maxwell equation in a non-homogeneous or random dielectric medium with non-linear response. I shall also discuss some approximation schemes, and prove in the random case a point-wise ergodic result. Details of the proofs and further results can be found in [1], [2]. In particular in [1] we prove that the formalism I use here can be adapted without essential changements to the study of other physical problems, e.g. thermal conductivity, magnetic susceptibility, elastic response, velocity of sound, viscous flow. In fact, all these problems are mathematically equivalent from the point of view considered here. The content of this contribution is the following. In Section 1 I describe the formalism in the case of an inhomogeneous medium and outline the proof that at least one solution exists when the non-linear terms are sufficiently small. In Section 2 I consider the case of a random medium, and show that it is mathematically equivalent to the one discussed in Section 1, apart from a difference which is pointed out. In Section 3 I discuss some properties of the effective dielectric constant and some approximation schemes. In Section 4, for a particular geometry and under an ergodicity assumption, I prove that the infinite-volume limit, for the electric field and for the effective dielectric constant, exist and are configuration independent.

In conclusion, I am glad to express here my thanks to Profs. S. Albeverio, Ph. Blanchard and L. Streit for the invitation to this Conference and for having provided a stimulating scientific atmosphere.

---

*) CNR, GNFM

1. Inhomogeneous Dielectric Media with Non-Linear Response

We denote by D a bounded domain in R^3, and by $\varepsilon(E)(x)$ a field of dielectric tensors. We assume that D be simply connected; this assumption can be easily lifted [1]. We write $\varepsilon(E)$ as

$$\varepsilon(E)(x) = \varepsilon^{(0)}(x) + \lambda\varphi(E)(x), \quad \lambda \geq 0 \; . \tag{1.1}$$

The parameter λ measures the strength of the non-linearity. We shall soon specify our assumptions on the dependence of φ on E. We want to find solutions of the equations

$$\left.\begin{array}{ll} \text{rot } E = 0 & (1.2)\text{a} \\ \text{div } \varepsilon(E)\cdot E = 0 & (1.2)\text{b} \end{array}\right\} \; x \in D$$

under suitable boundary conditions.

This will be achieved by generalizing the method of orthogonal projections, introduced in [4] and further developed in [2] for the linear case.

Since D is simply connected, equation (1.2)a implies that one can find a potential field ϕ such that $E = \nabla\phi$. Equation (1.2)b takes then the form

$$\text{div } \varepsilon(\nabla\phi) \cdot \nabla\phi = 0 \; . \tag{1.3}$$

Boundary conditions are defined by assigning a pair (V,ϕ_0) where V is a closed proper subspace of $H^1(D)$ such that $V \supseteq H^1_0$ and ϕ_0 is a pre-assigned element of $H^1(D)$ which does not belong to V. The spaces H^1_0 and H^1 are the standard Sobolev spaces. The boundary condition (V,ϕ) corresponds to

$$\phi - \phi_0 \doteq \psi \in V \; . \tag{1.4}$$

The function ϕ is a (weak) solution of (1.2) with boundary condition (V,ϕ_0) if there exists a function $\psi \in V$, such that $\phi = \phi_0 + \psi$ and ψ is a (weak) solution of

$$\text{div } \varepsilon(\nabla\phi_0 + \nabla\psi) \cdot \nabla(\phi_0 + \psi) = 0 \; . \tag{1.5}$$

In practical cases, one can choose ϕ_0 to be a solution in D of the equation $\Delta\phi_0 = 0$ and choose V to be the linear subspace of H^1 defined by boundary conditions of Dirichlet or mixed Dirichlet-Neumann type. The formalism allows of course for more general boundary conditions.

Define $T_\phi : V \to V^*$ through [3]

$$T_\phi(\psi)\cdot v = \langle\nabla v, \varepsilon(\nabla\phi_o + \nabla\psi)\cdot\nabla(\phi_o + \psi)\rangle - \langle\Delta v, \varepsilon(\nabla\phi_o)\cdot\nabla\phi_o\rangle \quad (1.6)$$

where $\langle,\rangle$ denotes the scalar product in $(L^2(D))^3$.

By definition, ψ is a weak solution of equation (1.5) iff

$$T_{\phi_o}(\psi) = f_o \quad (1.7)$$

where $f_o \doteq \mathrm{div}\ \varepsilon(\nabla\phi_o)\cdot\nabla\phi_o$.

Eq. (1.7) has at least one solution if T_{ϕ_o} is monotone and coercive [3]. In the present case, under the conditions on $\varphi(E)$ stated below, T_{ϕ_o} is coercive but not monotone. In general it is not variational; when it is variational for a specific choice of $\varphi(E)$, it does not satisfy a Palais-Smale condition. Therefore, the usual techniques do not apply.

We shall therefore proceed by adapting the approach of [4], which uses only a Hilbert-space structure. For this reason, it extends without change to the random case, as we shall see in Section 2. The price paid is that no regularity of E can be established. On the linear terms in (1.1) we make the assumption

A_0 There exist positive constants ε_+, ε_- such that

$$\varepsilon_- \cdot I \le \varepsilon^{(O)}(x) \le \varepsilon_+ \cdot I . \quad (1.8)$$

On the non-linear term in (1.1) we assume

A_1 For every $x \in D$ and $E \in (L^2(D))^3$, $\varphi(E)(x)$ is a symmetric tensor. Furthermore, there exists a function $h : R^+ \to R^+$, bounded on bounded intervals and infinitesimal at the origin, such that

$$\sup_{x\in D} |\varphi(E)(x)| \le h(\|E\|_2)$$

where $|\varphi(\cdot)|$ is the matrix norm and $\|E\|_2^2 \doteq \langle E,E\rangle$.

A_2 There exists a positive number α such that, for all $E \in (L^2(D))^3$

$$\varphi(E)(x) \geq \alpha \cdot\cdot I - \varepsilon_{\_} \cdot I .$$

Assumption A_2 together with A_o implies strong ellipticity of (1.2); it could be somewhat relaxed [1]

We remark briefly on A_1. This assumption requires that the dependence of φ on E be non-local, i.e. that $\varphi(E)(x_o)$ depend on the value of E in a full neighbourhood of the point x_o. Since the radius of non-locality can be chosen arbitrarily, assumption A_1 is rather mild from a physical point of view.

Assumptions A_1 and A_2 are satisfied for example by

$$\left.\begin{array}{l} \varphi(E)(x) = \frac{1}{|D|} \int_D g(x,x-y)E^2(y)dy \\ g \text{ bounded and positive.} \end{array}\right\} \qquad (1.10)$$

The support of g in the second variable will then be a measure of the amount of non-locality which is assumed. Our goal is to find solutions of equations $(1.2)_{a,b}$ with boundary condition (V,ϕ_o), under assumptions $A_o - A_2$.

We rewrite first equations $(1.2)_{a,b}$ using orthogonal projections. Let $G \doteq \nabla\phi_o$ and denote by H the smallest closed subspace of $(L^2(D))^3$ which contains all vector-valued (generalized) functions of the form ∇u, $u \in V$.

Let A be the orthogonal projections of $(L^2(D))^3$ onto H. We call it the Hodge projection, since it reflects the duality between the operators grad and div.

In this notation, equation (1.2)a and the boundary condition (v,ϕ_o) are together equivalent to the identity

$$A(E - G) = E - G . \qquad (1.11)$$

Define

$$B \doteq (1 - A)G . \qquad (1.12)$$

Since $\phi_o \notin V$, one has $AG \neq G$, and therefore $B \neq O$.

From (1.12) one derives

$$E - AE = B \,. \tag{1.13}$$

From $A_0 - A_2$ it follows that ε is a bounded map from H to $L^\infty(D,S)$, where S are symmetric matrices. Therefore $E \to \varepsilon(E) \cdot E$ is a (non-linear) bounded map from H to $(L^2(D))^3$ and (1.2)b can be written ([1], [4])

$$A\varepsilon(E) \cdot E = 0 \,. \tag{1.14}$$

We must find a solution of the system of equations (1.13), (1.14). To this purpose we shall write equations (1.13), (1.14) in an equivalent form. For every $\rho > 0$ define $\bar{\varepsilon}(\rho)$ through

$$\bar{\varepsilon}(\rho) \doteq \sup_{x\in D, \|E\|=\rho} |\varepsilon(E)(x)| \tag{1.15}$$

and remark that $\bar{\varepsilon}(\rho) \leq \varepsilon_+ + \lambda h(\rho)$.

Let $\varepsilon_0(\rho)$ be a positive-valued function greater than $\bar{\varepsilon}(\rho)$:

$$\varepsilon_0(\rho) \geq \bar{\varepsilon}(\rho) \quad \forall \rho \in R^+ \,. \tag{1.16}$$

Define

$$Q_{\varepsilon_0}(E)(x) \doteq \left(I - \frac{\varepsilon(E)(x)}{\varepsilon_0(\|E\|)}\right)^{1/2} \tag{1.17}$$

where one takes the positive square root. $Q_0(E)$ is a field of symmetric tensors of norm smaller than one.

One can then prove [2]

Lemma 1.1 Assume $A_0 - A_2$ and let $\bar{\varepsilon}$ be defined by (1.15). Then the system (1.13), (1.14) is equivalent to the equation

$$E = B + AQ_{\varepsilon_0}(I - Q_{\varepsilon_0} A\, Q_{\varepsilon_0})^{-1} Q_{\varepsilon_0} B \tag{1.18}$$

for each function $\varepsilon_0(\rho)$ which satisfies (1.16).

Another family of equations each of which is equivalent to (1.13), (1.14) can be obtained as follows. For each $\rho > 0$ define $\underline{\varepsilon}(\rho)$ through

$$\underline{\varepsilon}(\rho) = \inf_{x\in D, \|E\|=\rho} |\varepsilon(E)(x)| \tag{1.19}$$

and remark that $\underline{\varepsilon}(\rho) \geq \alpha$.

Let ε_0' be a strictly positive function smaller than $\underline{\varepsilon}(\rho)$:

$$0 < \varepsilon_0'(\rho) \leq \underline{\varepsilon}(\rho) \quad \forall \rho \in R^+ . \tag{1.20}$$

Define

$$Q'_{\varepsilon_0'}(E)(x) \doteq \left(\frac{\varepsilon(E)(x)}{\varepsilon_0'(\|E\|)} - I\right)^{1/2} . \tag{1.21}$$

Also $Q'_{\varepsilon_0'}$ is a field of symmetric tensors.

One can prove [2].

<u>Lemma 1.2</u> Assume $A_0 - A_2$ and let $\underline{\varepsilon}$ be defined by (1.19). Then the system (1.13), (1.14) is equivalent to the equation

$$E = B - AQ'_{\varepsilon_0'}(I + A'_{\varepsilon_0'} A Q'_{\varepsilon_0'})^{-1} Q'_{\varepsilon_0'} B \tag{1.22}$$

for each function ε_0' which satisfies (1.21).

In the remaining part of this section, $\varepsilon_0(\rho)(\varepsilon_0'(\rho))$ will be any function which satisfies (1.16) ((1.17)).

Let

$$b_\rho \doteq \{E \in H, \|E\| \leq \rho\} \tag{1.23}$$

and for each $\lambda \geq 0$ and $\rho \geq 0$ consider the map $\Phi_\lambda^{(1)} : b_\rho \to H$ defined by

$$\Phi_\lambda^{(1)}(W) \doteq A\, Q_0(1 - Q_0\, A\, Q_0)^{-1} Q_0 B \tag{1.24}$$

where

$$Q_0(W)(x) \doteq (I - \frac{\varepsilon(W+B)(x)}{\varepsilon_0(\|W+B\|)}) .$$

Since B is orthogonal to W, one has $\|B+W\|^2 = \|B\|^2 + \|W\|^2$. It is immediately seen from Lemma 1.1 that the solution of $(1.2)_{a,b}$ with boundary conditions (V, ϕ_0) are precisely the fixed points of the map $\Phi_\lambda^{(1)}$.

One can estimate easily the contractivity properties of $\Phi_\lambda^{(1)}$ since one has the explicit expression (1.24), which can be written also as a uniformly convergent series in the operator $Q_o A Q_o$.

One can then prove [2].

<u>Lemma 1.3</u> Under assumption $A_o - A_2$ one can find a positive constant λ_o depending on $\varepsilon^{(o)}$, h and B, with the following property: for $\lambda \in [0,\lambda_o]$ there exist non-empty intervals $I_i^\lambda \doteq \{\sigma | \underline{\sigma}_i^\lambda \leq \sigma \leq \sigma_i^{-\lambda}\}$, $i = 1,\dots,N_\lambda$ such that $\Phi_o^{(1)}$ maps b_ρ into itself if $\rho \in I_i^\lambda$ for some i .

Moreover, $\sigma_{N_\lambda}^{-\lambda} \to \infty$ and $\underline{\sigma}_1^\lambda \to a_o$ when $\lambda \to 0$, where a_o is the norm of the (unique) solution of the linear problem. An analogous result is obtained using Lemma 1.2.

For each $\lambda \geq 0$ and $\rho \geq 0$ define the map $\Phi_\lambda^{(2)} : b_\rho \to H$ by

$$\Phi_\lambda^{(2)}(W) \doteq -AQ_o'(I + Q_o' A Q_o')^{-1} Q_o' B \tag{1.25}$$

where

$$Q_o'(W) \doteq \left(\frac{\varepsilon(W+B)(x)}{\varepsilon_o'(\|W+B\|)} - I \right)^{1/2} . \tag{1.26}$$

From Lemma 1.2 it follows that the solutions of $(1.2)_{a,b}$ with boundary conditions (V,ϕ_o) are precisely the fixed points of the map $\Phi_\lambda^{(2)}$. Using the same estimates as in the proof of Lemma 1.3 one can then prove [2].

<u>Lemma 1.4</u> Under the assumptions $A_o - A_2$ the conclusion of Lemma 1.3 holds also for the map $\Phi_\lambda^{(2)}$. Moreover, the constant λ_o and the intervals I_i^λ coincide with those of the map $\Phi_\lambda^{(1)}$.

<u>Remark 1.5</u> Analogous results are obtained in [2] if one keeps λ fixed and considers the problem of invariant balls for sufficiently small values of $\|B\|$. One can in fact prove [2] that intervals with the property described in Lemma 1.3 exist for $\lambda \in [0,g(\|B\|)]$, where $g(\cdot)$ depends on $\varepsilon_\pm,\alpha$ and on $h(\rho)$, and $g(\sigma) \to \infty$ when $\sigma \to 0$.

Notice now that, when $\lambda = 0$

$$\Phi_o^{(1)}(W) = \Phi_o^{(2)}(W) = E_o - B$$

where E_o is the unique solution of the system $(2.1)_{a,b}$ in the linear case. Therefore, the fixed point is unique for $\lambda = 0$. For sufficiently small values of λ one may expect to have existence and possibly uniqueness.

Following [2] we give sufficient conditions for existence and for existence and uniqueness in a neighbourhood of E_o.

Consider the following assumptions

A_3 The map φ is Lifshitz-continuous in E uniformly over bounded sets in $H + B$.

Remark Assumption A_3 is satisfied by the example (1.10).

A_4 The map $E(\cdot) \to \varphi(E)(x)$ is continuous in x uniformly over bounded sets in $H + B$.

Remark Assumption A_4 is satisfied by (1.10) if $\sup_x |g(x,z)|$ is continuous.

One has then [2].

Proposition 1.6 Under assumptions $A_o - A_2$ and A_3 one can find a positive constant λ_1, and a function ρ_1 from $[0,\lambda_1)$ to R^+, depending on $\varepsilon_\pm$, B and $h(\rho)$ such that for all $\lambda \in [0,\lambda_1)$ the system $(1.2)_{a,b}$ has an unique solution in $b_{\rho_1(\lambda)}$.

The proof is obtained by proving that for sufficiently small λ there is a neighbourhood N_λ of E_o in which $\Phi_\lambda^{(1)}$ and $\Phi_\lambda^{(2)}$ are strict contractions. The positive constant λ_1 in Proposition 1.6 is in general considerably smaller than the constant λ_o which enters in Lemmas 1.4 and 1.5.

Assumption A_4 gives instead existence, but not necessarily uniqueness, for all $\lambda \in [0,\lambda_o]$. One has indeed [2]

Proposition 1.7 Under assumptions $A_o - A_2$ and A_4, the system $(1.2)_{a,b}$ with boundary conditions (V,ϕ_o) has at least one solution for every $\lambda \in [0,\lambda_o]$ where λ_o is the constant which appears in Lemma 1.3. One of the solutions has norm not greater than $((\underline{\sigma}_1^\lambda)^2 + |B|^2)^{1/2}$ where $\underline{\sigma}_1^\lambda$ is defined in Lemma 1.3.

The proof is given by verifying that, under Assumption A_4, the maps $\Phi_\lambda^{(1)}$ and $\Phi_\lambda^{(2)}$ are compact for every value of the parameter λ. Proposition 1.7 follows then from Lemma 1.4, Lemma 1.5 and the Leray-Schauder Lemma.

2. Random Media

In this section, we extend the formalism of Section 1 to the case or a random dielectric medium. To provide a motivation for the assumptions we shall make, we reformulate first the formalism of Section 1 for the case when the dielectric medium fills a unit cube and one imposes periodic boundary conditions.

We denote the unit cube by Ω, the Lebesgue measure by μ, and by ω a generic point in Ω. The pair (Ω,μ) defines a probability space on which the translation group $\mathbb{R}^3$ acts in a natural ergodic way:

$$\omega \to \omega + a \quad \text{mod } 1\ , \quad a \in \mathbb{R}^3\ .$$

If $E(\omega)$ is a measurable vector field on (Ω,μ), we construct a random field by

$$x \to E(x)(\omega) \doteq E(x+\omega)\ .$$

By construction $E(x)$ is strictly stationary (i.e. $E(x) \sim E(y)$ in distribution $\forall x,y \in R^3$) and one has

$$T_a E(x)(\omega) \doteq E(x)(\omega + a) = E(x+a,\omega)\ .$$

The representation $a \to T_a$ is strongly continuous and measure preserving. Let L_k be the infinitesimal generator of $T_{\hat{k}a}$, $a \in R$, $k=1,2,3$. Then L_k have a common dense core C and one has

$$L_k E(x) = \left(\frac{\partial}{\partial x_k} E\right)(x)$$

so that, in particular,

$$(A_{\text{periodic}})_{k\ell} = \frac{\partial}{\partial x_k} \cdot \Delta^{-1} \cdot \frac{\partial}{\partial x} = L_k \frac{1}{L^2} L_\ell\ .$$

Given the dielectric tensor field $\varepsilon(E)(\omega)$, one can construct then a stationary random field $\varepsilon(E)(x;\omega)$ by setting

$$T_{-x}\varepsilon(T_{-x}E)(x;\omega) = \varepsilon(E)(\omega) \quad \forall x \in R^3\ . \tag{2.1}$$

In the particular case (1.10) one has for example (with $g(x_1,x_2)$ independent of x_1)

$$\varepsilon(E)(x,\omega) = \varepsilon^{(0)}(x,\omega) + \lambda \int g(x-y)E^2(y,x)d^3y \cdot I \tag{2.2}$$

where

$$\varepsilon^{(0)}(E)(x,\omega) \doteq \varepsilon^{0}(x+\omega)\ .$$

Eq. (2.1) can be given the following form

$$\varepsilon(E)(\omega) \doteq \varepsilon^{(0)}(\omega) + \lambda \int g(-y)(T_y E)^2(\omega)dy \cdot I \,. \tag{2.3}$$

Maxwell's equation can be written

$$\begin{aligned} A_p(E-B) &= E-B \\ A_p \varepsilon(E)E(\omega) &= 0 \,. \end{aligned} \tag{2.4}$$

The system (2.4) can also be written

$$\begin{aligned} A_p(E(x,\cdot)-B) &= E(x,\cdot) - B \\ A_p \varepsilon(E(x,\cdot))\cdot E(x,\cdot) &= 0 \end{aligned} \tag{2.5}$$

where $E(x,\cdot)$ is the stationary random field associated to $E(\omega)$. The conditions $A_o - A_2$ can now be written

A_o' There exist positive constants $\varepsilon_\pm$ such that

$$\varepsilon_- \cdot I \le \varepsilon^{(0)}(\omega) \le \varepsilon_+ \cdot I \quad \forall \omega \in \Omega \,.$$

A_1' For every stationary random field $E(x)$, $\varphi(E)(x)$ is a s.s. random field with values in the symmetric tensors.

Furthermore, there exists a function $h : R^+ \to R^+$, bounded on bounded intervals and infinitesimal at the origin, such that

$$\sup_{\omega \in \Omega} |\varphi(E)(\omega)| \le h(\|E\|)$$

where

$$\|E\|^2 \doteq \int |E(\omega)|^2 \mu(d\omega) \,.$$

A_2' There exists a positive constant α such that, for all $E \in L^2(\Omega,\mu)$ one has $\varphi(E)(\omega) \ge (\alpha - \varepsilon_-)\cdot I \;\; \forall \omega \in \Omega$.

For a s.s. random field $Z(x,\omega)$ we have used the notation $Z(\omega) \doteq Z(0,\omega)$.

For the periodic case, conditions $A_o' - A_2'$ coincide with conditions $A_o - A_2$; therefore, the results of Section 1 can be given in this case a probabilistic version, by a simple rewording of Proposition 1.6 and 1.7.

We now want to obtain the same results in a general probabilistic setting under the assumptions $A_0' - A_2'$.

Remark 2.2 We shall prove the analogous of Proposition 1.6. Proposition 1.7 will in general not be true. In particular, we do not know how to characterize the triples (Ω,μ,T_a) which have the property that the map

$$E \in L^2(\Omega,\mu) \to \int g(-y)(T_y E)^2(\omega)dy \in L^2(\Omega,\mu)$$

is compact if g is continuous.

Let then (Ω,μ) be a probability space which carries a measure-preserving representation $a \to \tau_a$ of the translation group R^3 .

The points $\omega \in \Omega$ may represent translations, as in the periodic case and in the quasi-periodic one. In a more general context, Ω may be regarded as the collection of all possible realizations of the dielectric medium and then, for each ω, $\varepsilon(x)(\omega)$ is the realization map. The measure μ is any pre-assigned Borel probability measure on the space of realizations, with the only condition to be invariant and ergodic under the natural action of the translation group.

For every measurable function F we define

$$T_a F(\omega) \doteq F(\tau_a \omega) \qquad a \in R^3 . \tag{2.6}$$

A strictly stationary (s.s) random tensor field Z is then a measurable map from $\Omega \times R^3$ to a suitable vector space, with the property

$$T_a Z(x) \simeq Z(x+a)$$

where $\simeq$ stands for identity in distribution.

We shall use the notation

$$Z(\omega) \doteq Z(0,\omega) .$$

The correspondence $Z(x,\cdot) \to Z(\omega)$ preserves all algebraic relations and the functional calculus. We shall therefore state assumptions and results in terms of the random variable $Z(\omega)$.

With the notations introduced so far, Maxwell's equations read

$$\left.\begin{aligned} &\mathrm{rot}_x\, E(x,\omega) = 0 \\ &\mathrm{div}\, \varepsilon(E)(x,\omega)E(x,\omega) = 0 \end{aligned}\right\} \quad \mu\text{-almost all } \omega . \tag{2.7}$$

We solve (2.7) under the following "boundary" condition

$$\int E(x,\omega)\mu(d\omega) = B \tag{2.8}$$

where B is a pre-assigned constant vector.

Notice that, by stationarity, the left hand side of (2.8) is independent of x. More general boundary conditions can be introduced ([2]); we shall not discuss here this point.

Denote by L_k, $k = 1,2,3$ the generators of $T_{a\hat{k}}$, $a \in R$, where $\hat{1},\hat{2},\hat{3}$ are three orthogonal unit vectors.

By a theorem of Stone and Kadison, the operators L_k have a dense common core G. On G^3 we define $\underset{\sim}{A}_p$ by

$$(\underset{\sim}{A}_p f)_k(\omega) \doteq L_k \cdot L^{-2}(L\cdot f)(\omega) \tag{2.9}$$

where $L^2 = L_k L_k$.

It follows from ergodicity that the kernel of L^2 is spanned by constant (vector-valued) functions.

It is then straightforward to verify that $\underset{\sim}{A}_p$ is closable and uniformly bounded by I, and that its unique extension to $(L^2(\Omega,\mu))^3$ is a projection operator, which we denote by A_p. The operator A_p commutes with T_a for all $a \in R^3$ and leaves G^3 invariant. We denote by H_p its range.

One can prove [2].

<u>Lemma 2.3</u> The system (2.7) together with the boundary condition (2.8) is equivalent to

$$\begin{aligned} A_p(E-B)(\omega) &= E(\omega) - B(\omega) \\ A_p\varepsilon(E)\, E(\omega) &= 0 . \end{aligned} \tag{2.10}$$

Sketch of proof: Define $H^1(\Omega)$ through

$$H^1(\Omega) = \{\varphi \in L^2(\Omega,\mu),\quad \varphi \in \text{Dom}\, L_k,\ k = 1,2,3\} .$$

The closure is taken in the Hilbert norm

$$\|\varphi\|_1 \doteq \{\int |\varphi(\omega)|^2\mu(d\omega) + \int \sum_{k=1}^{3} |L_k\varphi(\omega)|^2\mu(d\omega)\}^{1/2} .$$

We denote by $H^{-1}(\Omega)$ the dual of $H^1(\Omega)$ with respect to the scalar product in $L^2(\Omega,\mu)$.

Using the invariance of G under T_a one verifies that

$$(\frac{\partial}{\partial x_j}\ \varepsilon_{jm}(E)E_m)(x,\cdot) \in H^{-1}(\Omega) \qquad \forall x \in R^3$$

and also that the second equation in (2.7) is equivalent to

$$\int (L_j\varphi)(\omega)(\varepsilon_{jm}(E)E_m)(\omega)\mu(d\omega) = 0 \qquad \forall \varphi \in H^1(\ ) \tag{2.12}$$

which in turn is equivalent to the second equation in (2.10). In the same way one verifies that, in $H^{-1}(\Omega)$

$$\mathrm{rot}_x E(x,\omega) = 0 \Leftrightarrow (L \wedge E)(\omega) = 0 \ . \tag{2.13}$$

It is then straightforward to verify the equivalence of the first two equations in (2.10) and (2.13), respectively. □

We now notice that the system (2.10) is identical in form to the system (1.13), (1.14). Therefore, the approach of Section 1 can be extended to the random case and one has

<u>Proposition 2.4</u> Assume $A_0' - A_2'$. The system (2.7) with boundary condition (2.8) is equivalent to

$$E = B + A_p Q_0 (1 - Q_0 A Q_0)^{-1} Q_0 B \tag{2.14}$$

for any function $\varepsilon_0(\rho)$ which satisfies (1.16), and it is also equivalent to

$$E = B - A_p Q_0' (1 + Q_0' A Q_0')^{-1} Q_0' B \tag{2.15}$$

for any function $\varepsilon_0'(\rho)$ which satisfies (1.21). The random variables Q_0 and Q_0' are defined by

$$Q_0(\omega) \doteq \left(I - \frac{\varepsilon(E)(\omega)}{\varepsilon_0(\|E\|)}\right)^{1/2}, \quad Q_0'(\omega) \doteq \left(\frac{\varepsilon(E)(\omega)}{\varepsilon_0'(\|E\|)} - I\right)^{1/2} .$$

From Proposition 2.4 it follows that the results of Section 1 hold also in the random case, with the possible exception of Proposition 1.7 (see Remark 2.2).

We conclude this section with a brief comment on the relation between the results of Sections 1 and 2. This will also provide a motivation for the problem discussed in Section 4.

One can write (2.14) in the form

$$E(x,\omega) = B + A_p Q_o(x,\omega)(I - Q_o(x,\omega)A_p Q_o(x,\omega))^{-1} Q_o(x,\omega)B \ . \qquad (2.16)$$

Comparing with (1.18), one is led to inquire whether one can find for μ-a.a.ω a projection operator $A(\omega)$ (possibly depending on ω) such that (2.15) can be written in the form (1.18), ω playing now the role of parameter. This would mean that $E(x,\omega)$ is a solution of Maxwell's equations when the dielectric medium has configuration ω.

"Deterministic" Hodge projections $A(\omega)$ with this property do not exist, in general. A notable exception is the periodic case, as remarked at the beginning of this section. Another exception occurs when all realizations of the dielectric medium have the form

$$\varepsilon(x,\omega) = \varepsilon^{(o)} + \varepsilon^{(2)}(E)(x,\omega) \ , \quad x \in R^3 \ , \quad \omega \in \Omega \qquad (2.17)$$

where $\varepsilon^{(o)}$ is a given symmetric tensor and $\varepsilon^{(2)}$ is such that, for each $i,j = 1\ldots3$, the expression

$$\int \varepsilon_{ij}^{(2)}(E)(x,\omega)d^3x$$

defines a bounded map from $(L^2(\Omega,\mu))^3$ to $L^\infty(\Omega,\mu)$.

Consider the Hilbert space

$$(L^2(R^3))^3 \oplus \underset{\sim}{R}^3$$

where $\underset{\sim}{R}$ is the space of constant functions.

One can prove that $A(\omega)$ exists for μ-a.a.ω, is independent of ω, has R^3 as kernel and coincides in $(L^2(R^3))^3$ with the Hodge projection there.

This case is, however, of little physical interest, since for every $E \in (L^2(\Omega,\mu))^3$ and μ-a.a.ω one has

$$\lim_{L\to\infty} \frac{1}{L^3} \int_{V_L} |\varepsilon^{(2)}(E)(x,\omega)|^2 dx = 0 \ .$$

Here V_L is a cube of size L centred at the origin.

Therefore, all the realizations of the medium appear homogeneous and with linear response at very large scales.

In significant physical situations one expects that the dielectric medium appears inhomogeneous at all scales.

One must then consider the infinite-volume limit to provide a connection between the random field $E(x)$ described in Section 2 and the field $E(x,\omega)$ constructed in Section 1 when the dielectric medium is in a state ω. Techniques and results are then typical of ergodic theory. This problem will be discussed in Section 4.

3. Average Properties and an Asymptotic Expansion

In this section we shall study average properties of some function of the elastic field E, and in particular of the electric displacement field $\underset{\sim}{D} \doteq \varepsilon(E)E$.

In the deterministic case, the average is taken with respect to Lebesgue measure over the domain D occupied by the dielectric medium. In the random case, the average is taken with respect to the measure μ over the probability space Ω. In both cases, $\bar{\xi}$ will denote the average of ξ. We shall consider only boundary conditions such that B is constant. It is then easy to verify that

$$\overline{E \cdot B} = |B|^2 .$$

The effective dielectric constant is by definition the average of $\underset{\sim}{D}$ in the direction of B, divided by the average of E in the same direction

$$\varepsilon^* \doteq \frac{\overline{\underset{\sim}{D}\cdot B}}{\overline{E\cdot B}} = \frac{1}{|B|^2}\;\overline{\underset{\sim}{D}\cdot B} \quad . \tag{3.1}$$

When ε does not depend on E, this definition is suggested by linear response theory and it turns out that ε^* is independent of $|B|$. In the non-linear case, ε^* will depend in general on $\overline{E\cdot B}$, and provide a description of an effective non-linear response.

An analytic expression is obtained easily from (1.18), (1.22). When $\lambda = 0$, one can take $\varepsilon_0(\rho) = \varepsilon_0 \;\; \forall \rho$, so that Q_0 is a given tensor field and (1.18) provides an explicit expression for ε^* in terms of the spectral properties of the operator $Q_{\varepsilon_0} A Q_{\varepsilon_0}$. When $\lambda > 0$ we write

$$Q_{\varepsilon_0}(\lambda) = Q_c(0) + \lambda Z(\lambda) \tag{3.2}$$

where $c = \varepsilon_0(\|E_0\|)$ and E_0 is the unique solution of $(1.2)_{a,b}$ for $\lambda = 0$.

From the first resolvent identity one has

$$(I - Q_{\varepsilon_o}(\lambda)AQ_{\varepsilon_o}(\lambda))^{-1} = (I - Q_c AQ_c)^{-1} + o(\lambda). \tag{3.3}$$

From (1.18) one can now determine E_1 in the following expansion

$$E(\lambda) = E_o + \lambda E_1 + o(\lambda) . \tag{3.4}$$

If the functional $\varphi(E)$ is of class C^K, this procedure can be iterated k times; in particular, if $\varphi(E)$ is of class C^∞ (e.g. if it depends polynomially on E) the iteration can be continued indefinitely. One proves then [2].

<u>Lemma 3.1</u> If $\varphi(E)$ is of class C^∞, the solution E of Maxwell's equation, described in Proposition 1.6, has an asymptotic expansion in powers of λ, i.e. there are functionals R_i, $i = 1,2,\ldots,$ such that for all $M \geq 1$

$$E(\lambda) = E_o + \sum_{m=1}^{M} \lambda^m R_m(E_o) + o(\lambda^M) . \tag{3.5}$$

The first new functionals R_i are constructed in [2]. From Lemma 3.1 one derives easily the corollary:

<u>Corollary 3.2</u> Under the assumption of Lemma 3.1, the effective dielectric constant ε^* has an asymptotic expansion in λ, i.e. there are constants c_n, $n \geq 1$ which depend only on B,A and $\varepsilon^{(o)}$ such that, for every integer $N \geq 1$

$$\varepsilon^*(\lambda) = \varepsilon^*(0) + \sum_{k=1}^{N} c_n \lambda^n + o(\lambda^N) . \tag{3.6}$$

An important special case is obtained if $\varepsilon^{(o)}$ takes only a finite number of values $\varepsilon_1 \ldots \varepsilon_M$. Consider for definiteness the case $\varepsilon_m = x_m \cdot I$. One can prove [1], [2] that $\varepsilon^*(0)$ and all the c_n's in (3.6) are boundary values of functions

$$\varepsilon^*(0)(z_1 \ldots z_M), \quad c_n(z_1 \ldots z_M) \qquad n = 1,\ldots$$

which are analytic in a domain Θ defined as follows. Θ is the smallest connected domain in C^M with the following properties

a) Θ contains the point $\{z | z_i = 1, \quad i = 1\ldots M\}$.

b) The boundary $\partial\Theta$ is contained in the union of the "half hyperplanes" Γ_{ij}, $i \neq j = 1\ldots M$ defined by

$$\Gamma_{ij} = \{z|z_i\bar{z}_j \in R^-\} .$$

This proof makes use of the structure of Equations (1.18) and (1.22). For $\varepsilon^*(0)$ the result follows also directly from the elliptic character of $(1.2)_{a,b}$.

Using Weyl's theory of analytic polyhedra, one can give a representation for functions analytic in Θ. This representation together with (1.18) could be used to provide estimates on ε^* when its value is known for some special choices of $x_1 \dots x_M$.

4. The Infinite Volume Limit

In this section we shall study the relation between the results of Section 1 and those of Section 2. We will do this for a special choice of domain in R^3 ; more general cases can be treated similarly.

We shall follow closely the method of proof in ([5], Appendix); the results we obtain are stronger, even in the linear case.

The problem is the following: as in Section 2, let Ω be the collection of all possible realizations of the dielectric medium. All realizations are assumed to be periodic of period one along two orthogonal directions in R^3, so that they can be viewed as tensor-valued fields on $T^2 xR$. We denote by (x_1, x_2, z) a generic point of $T^2 xR$.

Each realization ω provides then a map

$$E \to \varepsilon^{(o)} + \lambda\varphi(E)$$

from vector fields on $T^2 xR$ to symmetric tensor fields on the same domain. We assume as given a Borel probability measure μ on Ω, invariant and ergodic under the natural action τ_a of the translational group on Ω. If S is a stationary random field of mean zero, which for μ-almost all ω is locally integrable in x , we define

$$\Omega_S \doteq \{\omega \in \Omega | \lim_{L\to\infty} \frac{1}{2L} \int_{-L}^{L} S(x_1, x_2, z; \omega) ds = 0 \text{ in } L^2(T^2)\}. \quad (4.1)$$

We shall say that (τ_a, μ) is ergodic if the following assumption is satisfied

α_1 : for all S of the class described above, $\mu(\Omega_S) = 1$.

We shall also need an assumption on the dependence of φ on E ;

in the linear case this assumption is of course trivially satisfied. We use the same notation as in Section 2, and take $B = b\hat{k}$, where $\hat{k}$ is the unit vector in the z-direction.

For given $\omega \in \Omega$ and $L > 0$, let $E_L(x,\omega)$ be the solution of $(1.2)_{a,b}$ in $T^2x[-L,L]$ with dielectric tensor $\varepsilon(E)(x,\omega)$ and boundary conditions

$$E_L(x,\omega) - b\hat{k} = \nabla U, \quad U(L) = U(-L) = 0 . \tag{4.2}$$

Denote by $E(x,\omega)$ the s.s. random field solution of (2.7) with boundary condition

$$\int E(x,\omega)\mu(d\omega) = b\hat{k} . \tag{4.3}$$

With this notation, the assumption we make on $\varepsilon(E)$ is the following

α_2: There exists a function $\gamma: R^+ \times R^+ \to R^+$, infinitesimal at the origin in the second variable, such that

$$\sup_L \sup_{x\in V_L} |\varepsilon(E)(x,\omega) - \varepsilon(E_L)(x,\omega)| < \gamma(\|E(\cdot,\omega) - E_L(\cdot,\omega)\|_L, \lambda) \tag{4.4}$$

where

$$\|f(\cdot)\|_L^2 = \frac{1}{2L}\int_{V_L} |f(x)|^2 dx , \quad V_L \doteq T^2x[-L,L] . \tag{4.5}$$

One can then prove

<u>Proposition 4.1</u> Assume α_1 and α_2. There exists then a subset $\Omega_o \subset \Omega$ such that $\mu(\Omega_o) = 1$ and if $\omega \in \Omega_o$

$$\lim_{L\to\infty} \frac{1}{2L}\int_{V_L} |E(x,\omega) - E_L(x,\omega)|^2 dx = 0 . \tag{4.6}$$

<u>Proof</u> Let

$$\Omega_1 \doteq \{\omega|\ \lim_{L\to\infty} \frac{1}{2L}\int_{-L}^{L}(E(\underline{x},z;\omega) - b\hat{k})dz = 0 \ \text{ in } \ L^2(T^2)\} . \tag{4.7}$$

By assumption α_1, $\mu(\Omega_1) = 1$.

For $\omega \in \Omega_1$ define

$$U(\underline{x},z;\omega) \doteq \int_{-z}^{z}(\hat{k}\cdot E(\underline{x},s;\omega) - b)ds + \int_0^1 \hat{\xi}\cdot E(r\hat{\xi},0;\omega)dr \tag{4.8}$$

where $\hat{\xi} = (\frac{x_1}{|\underline{x}|}, \frac{x_2}{|\underline{x}|})$ and $\underline{x} = (x_1, x_2)$.

It is straightforward to verify that

$$\nabla U = E - b\hat{k} . \tag{4.9}$$

Moreover, when $\omega \in \Omega$

$$\frac{1}{L} U(\underline{x}, L; \omega) \to 0 \quad \text{in} \quad L^2(T^2) .$$

Let

$$\beta(L,\omega) \doteq \int_{T^2} U^2(\underline{x}, L; \omega)\, dx \tag{4.10}$$

and remark that

$$\beta(L,\omega) \xrightarrow[L \to \infty]{} 0 \quad \text{if} \quad \omega \in \Omega_1 . \tag{4.11}$$

We define Ω_2 as follows

$$\Omega_2 = \{\omega | \lim_{L\to\infty} \frac{1}{2L} \int_{-L}^{L} [E^2(\underline{x}, s; \omega) - \int E^2(\omega)\mu(d\omega)]ds = 0 \quad \text{in} \quad L^2(T^2)\} . \tag{4.12}$$

Let $\Omega_o = \Omega_1 \cap \Omega_1$; by assumption $\mu(\Omega_o) = 1$.

Let $\omega \in \Omega_o$; we shall prove that one can find [5] a sequence of domains $V_N(\omega)$:

$$V_N(\omega) = T^2 x [-L_N(\omega), + L_N(\omega)]$$

and a sequence of functions $\chi_N(z,\omega)$ of class C^1 in the first variable, with the following property

a) $|\chi_N| \leq 1$, $\chi_N(z,\omega) = 0$ if $|z| \geq L_N(\omega)$

b) $\frac{1}{|V_N|} \int_{V_N} (1 - \chi_N) E^2(x,\omega)\, dx \xrightarrow[N \to \infty]{} 0$ (4.13)

c) $\frac{1}{|V_N|} \int_{V_N} (\nabla\chi_N)^2\, U^2(x,\omega)\, dx \xrightarrow[N \to \infty]{} 0$.

Assuming for the moment that such sequence of functions has been constructed, we complete the proof of Proposition 4.1.

We shall treat separately the linear and non-linear case.

1) Linear case.

Let U_L be a potential for $E_L(x,\omega) - b\hat{k}$, with $U_L(0,\omega) = 0$. One has

$$\|E_L - E\|_L^2 = \frac{1}{2L} \int |E_L(x,\omega) - E(x,\omega)|^2 dx \le$$

$$\le \frac{1}{\varepsilon_-} \frac{1}{2L} \int_{V_L} [(\nabla U - \nabla U_L)\varepsilon^0(\nabla U_L - \nabla\chi_L U - U\nabla(1-\chi_L)]dx \le \tag{4.14}$$

$$\le \frac{\varepsilon_+}{\varepsilon_-} \frac{1}{2L} \int_{V_L} |\nabla U_L - \nabla U|\,|(1-\chi_L)\nabla U + U\nabla\chi_L|dx \,.$$

In deriving (4.14) we have used $(1.2)_{a,b}$, (2.7) and the inequality

$$\varepsilon_- \cdot I \le \varepsilon^{(o)}(x,\omega) \le \varepsilon_+ \cdot I \,.$$

From Schwartz's inequality one has then

$$\|E_L - E\|_L^2 \le \frac{\varepsilon_+}{\varepsilon_-} \frac{2}{L} \left[\int |U^2(\nabla\chi_L)^2 dx + \int (1-\chi_L)(\nabla U)^2 dx \right] \tag{4.15}$$

so that, using a), b), c), if $\omega \in \Omega_0$

$$\|E_L(\cdot,\omega) - E(\cdot,\omega)\|_L^2 \to 0 \,.$$

2) Non-linear case.

Proceeding as in case 1), one obtains

$$\|E_L(\cdot,\omega) - E(\cdot,\omega)\|_L^2 \le \frac{\varepsilon_+}{\varepsilon_-} \frac{1}{2L} \int_{V_L} |\nabla U_L - \nabla U|\,|\nabla U(1-\chi_L) + U\nabla\chi_L|dx + \tag{4.16}$$

$$+ \frac{1}{\varepsilon_-} \frac{1}{2L} \int_{V_L} |\nabla U\cdot(\varepsilon(E) - \varepsilon(E_L))\nabla(U - U_L)|dx \,.$$

From (1.1) one has

$$\varepsilon(E) - \varepsilon(E_L) = \lambda(\varphi(E) - \varphi(E_L))$$

so that

$$\left|\frac{1}{\varepsilon_-}\frac{1}{2L}\int_{V_L} |\nabla U\cdot(\varepsilon(E)-\varepsilon(E_L))\nabla(U-U_L)|dx \le \lambda\|E\|\,\|E_L-E\|_L\ \|\varphi(E)-\varphi(E_L)\|_\infty .\qquad (4.17)$$

From (4.16) and (4.17) together with assumption α_2 it follows

$$(1-\|E\|\gamma(\|E-E_L\|_L,\lambda))\|E-E_L\|_L \le \qquad (4.18)$$

$$\frac{\varepsilon_+}{\varepsilon_-}\left[\frac{1}{2L}\int_{V_L}(\nabla U)^2(1-\chi_L)^2dx + \frac{1}{2L}\int_{V_L}\phi^2(\nabla\chi_L)^2dx\right]^{1/2} .$$

From assumption α_1 and the definition of the domain Ω_0 it follows that the right hand side of (4.18) converges to zero when $L\to\infty$. Moreover, one can find a positive constant λ_2 (smaller than the constant λ_0 in Proposition 1.7) such that

$$\lambda\|E\|\gamma(\|E-E_L\|_L,\lambda) < 1 \quad \text{for} \quad \lambda\in[0,\lambda_2)$$

(notice that E and E_L depend on λ, but Proposition 1.7 provides an upper bound on $\|E\|$ and $\|E_L\|_L$ for $\lambda\in[0,\lambda_1)$.

We conclude that, if $\lambda\in[0,\lambda_2]$ and $\omega\in\Omega_0$

$$\|E(\cdot,\omega) - E_L(\cdot,\omega)\|_L \to 0 \quad \text{when} \quad L\to\infty .$$

It remains therefore to prove that one can construct ([5], Appendix) a sequence of functions χ_N with the properties a,b,c described in (4.13).

For each $\omega\in\Omega_0$, we define $\tilde{\chi}_L(s,\omega)$ by

$$\tilde{\chi}_L(s,\omega) = 0 \qquad s=\pm L$$

$$\tilde{\chi}_L(s,\omega) = 1 - \frac{s}{L(1-\beta(L,\omega))} , \quad L(1-\beta(L,\omega)) \le |s| \le L \qquad (4.19)$$

$$\tilde{\chi}_L(s,\omega) = 1 \qquad -L(1-\beta(L,\omega)) \le s \le L(1-\beta(L,\omega)) .$$

Let $\approx\chi_L$ of class C^∞ be obtained by rounding off $\tilde{\chi}(s)$ in a neighbourhood of $\pm L$ and of $\pm L(1-\beta(L,\omega))$.

Define $\chi_L(\underline{x},s;\omega)$ by

$$\chi_L(\underline{x},s;\omega) \doteq \tilde{\tilde{\chi}}_L(s,\omega) \ . \tag{4.20}$$

It is straightforward to verify that χ_L satisfies a), b), c) in (4.13). This concludes the proof of Proposition 4.1.

From Proposition 4.1 one derives an ergodic result on the effective dielectric constant. We formulate it as

Proposition 4.2 Let $\varepsilon_L^*(\omega)$ be the effective dielectric constant for $E_L(x,\omega)$ and let ε^* be the effective dielectric constant of the random case. One can find a set $\Omega_3 \subset \Omega$ with the properties

$$\mu(\Omega_2) = 1 \ ; \ \lim_{L\to\infty} \varepsilon_L^*(\omega) = \varepsilon^* \quad \forall\omega \in \Omega_3 \ .$$

Proof Define Ω_4 by

$$\Omega_4 = \{\omega | \lim_{L\to\infty} \frac{1}{2L} \int_{-L}^{L} \left[(T_{s\hat{k}}\, \varepsilon(E)E)(\omega) - \int (\varepsilon(E)E)(\omega)\mu(d\omega) \right] ds = 0\} \ . \tag{4.21}$$

Let

$$\Omega_3 \doteq \Omega_0 \cap \Omega_4$$

and notice that

$$\mu(\Omega_3) = 1$$

in view of assumption α_1 .

Define

$$\underset{\sim}{\varepsilon}_L^*(\omega) \doteq \frac{1}{2L} \int_{V_L} (E(x,\omega)\varepsilon(E)(x,\omega)\hat{k})dx \ . \tag{4.22}$$

Since $\Omega_3 \subset \Omega_4$, one has

$$\lim_{L\to\infty} (\varepsilon_L^*(\omega) - \varepsilon^*) = 0 \quad \forall\omega \in \Omega_3 \ .$$

Let

$$c_1 \doteq \sup_{\omega} |\varepsilon(E)(\omega)| \ , \quad c_2 \doteq \sup_{L} ||E_L||_L \ .$$

Notice that $c_2 < \infty$ since $\|E_L - E\|_L \to 0$ and $\omega \in \Omega$.
Since $\Omega_3 \subset \Omega_0$,

$$\lim_{L\to\infty} |\varepsilon_L^*(\omega) - \varepsilon_L^*(\omega)| \leq$$

$$\leq \lim_{L\to\infty} \frac{1}{2L} \int_{V_L} |((E(x,\omega) - E_L(x,\omega))\varepsilon(E)(x,\omega)\cdot\hat{k})|dx \tag{4.24}$$

$$+ \lim_{L\to\infty} \frac{1}{2L} \int_{V_L} |(E_L(x,\omega)(\varepsilon(E) - \varepsilon(E_L))(x,\omega)\cdot\hat{k})|dx$$

so that, using Theorem 4.1 and Assumption A_2'

$$\lim_{L\to\infty} |\varepsilon_L^*(\omega) - \underset{\sim}{\varepsilon}_L^*(\omega)| = 0 \quad \forall\omega \in \Omega_3 . \tag{4.25}$$

Proposition 4.2 follows now from (4.23) and (4.25). □

References

[1] G.F. DELL'ANTONIO, R. FIGARI, E. ORLANDI, Representation and asymptotic expansion for the electric field in random or inhomogeneous media. University of Marseille preprint, June 1984.

[2] G.F. DELL'ANTONIO, Existence and representation theorems for non-linear electrostatics in inhomogeneous media, ZiF, University of Bielefeld preprint, October 1984.

[3] F. GILBARG, N. TRUDINGER, Elliptic PDE of second order, Springer Verlag Berlin-Heidelberg-New York 1977.

[4] H. WEYL, Duke Math. Journal 7, (1940).

[5] K. GOLDEN, G. PAPANICOLAU, Comm. Math. Phys. 90, 473 (1983).

REMARKS ON THE CENTRAL LIMIT THEOREM FOR WEAKLY DEPENDENT RANDOM VARIABLES

Detlef Dürr*, Sheldon Goldstein§
Institut des Hautes Etudes Scientifiques
91440 Bures-sur-Yvette (France)

I. Introduction

In recent years the Central Limit Theorem (CLT) for martingale differences [1,2] has been frequently applied to stationary sequences $\{X_i\}_{i\in\mathbb{Z}}$ to establish the convergence of the normalized sums

$$\frac{1}{\sqrt{n}} S_n = \frac{1}{\sqrt{n}} \sum_{i=0}^{n-1} X_i$$

to a normal law. A sequence $\{m_i\}_{i\in\mathbb{Z}}$ of random variables adapted to some increasing family of σ-algebras $\{\mathcal{F}_i\}_{i\in\mathbb{Z}}$ are called martingale differences if $E(m_{i+1}|\mathcal{F}_i) = 0$ for all i.

The first one to observe that the standard results for CLT's for weakly dependent random variables may be obtained from the CLT for martingale differences was Gordin [3]. This approach was then taken up by others (see the monograph by Hall and Heyde and references therein [4]).

The idea is to approximate X_i by a martingale difference m_i in such a way that the "error"

$$\frac{1}{\sqrt{n}} \sum_{i=0}^{n-1} (X_i - m_i) \to 0$$

in probability. Then one obtains easily that $S_n/\sqrt{n}$ converges to the same law to which $\frac{1}{\sqrt{n}} \sum_{i=0}^{n-1} m_i$ converges.

In these notes we derive, using a naive and simple approach, very weak conditions under which a stationary ergodic sequence may be approximated by a stationary ergodic sequence of martingale differences

* Supported by a DFG-Heisenberg Stipend, present address: BIBOS, Universität Bielefeld, 4800 Bielefeld, West-Germany

§ Research partially supported by NSF Grant N° PHY-8201708; on leave from Department of Mathematics, Rutgers University, New Brunswick, New Jersey 08903.

From this we obtain immediately the classical CLT's for weakly dependent random variables. For completeness we review in the next section the classical results. Section III deals with the approximation and contains our main result, a direct consequence of a simple (abstract) Hilbert space lemma, proven at the end of this paper in section V. In section IV we generalize the approximation procedure to situations which frequently appear in applications, namely when X_i is not measurable w.r.t. the σ-algebra $\mathcal{F}_i$, which is the one used to define the martingale difference m_i.

II. Classical results

Let $\{X_i\}_{i\in\mathbb{Z}}$ be a stationary sequence of centered square integrable random variables defined on a probability space $(\Omega,\mathcal{F},\mathbb{P})$.

The classical conditions under which a CLT for the normalized sum

(2.1) $$\frac{1}{\sqrt{n}} S_n = \frac{1}{\sqrt{n}} \sum_{i=0}^{n-1} X_i$$

holds, are in the form of mixing conditions (weak dependence of past and future) which enables one to reduce the problem to a CLT for independent random variables:

Let
$$\mathcal{F}_0 = \sigma(X_j, j \leq 0)$$
denote the σ-algebra of the past up to time 0 and
$$\mathcal{F}^n = \sigma(X_j, j \geq n)$$
that of the future after time n.

Two standard mixing properties are

ϕ-mixing: $$\phi(n) = \sup_{\substack{G\in\mathcal{F}^n \\ F\in\mathcal{F}_0, \mathbb{P}(F)>0}} |\mathbb{P}(G|F) - \mathbb{P}(G)| \underset{n\to\infty}{\to} 0$$

and

α-mixing: $$\alpha(n) = \sup_{\substack{G\in\mathcal{F}^n \\ F\in\mathcal{F}_0}} |\mathbb{P}(GF) - \mathbb{P}(G)\,\mathbb{P}(F)| \underset{n\to\infty}{\to} 0$$

Equivalently ϕ and α may be expressed in terms of expectations of $\mathcal{F}^n$-measurable functions $h(h\in\mathcal{F}^n)$ and $\mathcal{F}_0$-measurable functions $f(f\in\mathcal{F}_0)$:

$$\phi(n) = \frac{1}{2} \sup_{\substack{|f|\leq 1 \\ |h|\leq 1}} \frac{|E(fh) - E(f)E(h)|}{E(|f|)}$$

and

$$\alpha(n) = \frac{1}{4} \sup_{\substack{|f|\leqslant 1 \\ |h|\leqslant 1}} |E(fh) - E(f)E(h)|$$

Another mixing condition involving the correlation is

ρ-mixing: $$\rho(n) = \sup_{\substack{\|f\|_2\leqslant 1 \\ \|h\|_2\leqslant 1}} \frac{E(fh) - E(f)E(h)}{\|f\|_2 \ \|h\|_2} \underset{n\to\infty}{\to} 0 \quad ; \ f\in\mathfrak{F}_0 \ , \ h\in\mathfrak{F}^n$$

where $\| \ \|_2$ denotes the norm of $L_2(d\,\mathbb{P})$.

For later use we state estimates of certain conditional expectations which are simple consequences ofthe above definitions [4]:

(2.2) $\|E(h|\mathfrak{F}_0)\|_2 \leqslant \rho(n)\|h\|_2$

(2.3) $\|E(h|\mathfrak{F}_0)\|_2 \leqslant 2\ \phi(n)^{1-1/r} \ \|h\|_r \quad , \ 2 \leqslant r \leqslant \infty$

(2.4) $\|E(h|\mathfrak{F}_0)\|_u \leqslant 2(2^{\frac{1}{u}} + 1)\ \alpha(n)^{\frac{1}{u}-\frac{1}{r}} \ \|h\|_r \quad , \ 1 \leqslant u \leqslant r \leqslant\infty$

where $\| \ \|_p$ denotes the $L_p(d\,\mathbb{P})$ norm.

<u>(2.5) Theorem</u> (Ibragimov, Linnik [5,6]):
Suppose that $\{X_i\}_{i\in\mathbb{Z}}$ is a stationary ρ, respectively ϕ, respectively α-mixing sequence. If $E(|X_0|^{2+\delta}) < \infty$ and

(2.6) $$\sum_{n=0}^{\infty} \rho(n) < \infty \ , \ \delta \geqslant 0$$

respectively

(2.7) $$\sum_{n=0}^{\infty} \phi(n)^{\frac{1+\delta}{2+\delta}} < \infty \ , \ \delta \geqslant 0$$

respectively

(2.8) $$\sum_{n=0}^{\infty} \alpha(n)^{\frac{\delta}{2+\delta}} < \infty \ , \delta > 0$$

then

$$\frac{1}{\sqrt{n}} S_n \Rightarrow \mathcal{N}(0,\sigma^2) \quad , \text{ as } n \to \infty$$

where

(2.9) $$\sigma^2 = \sum_{n=-\infty}^{\infty} E(X_n X_0) \quad . \quad \square$$

Here " $\Rightarrow$ " denotes the convergence in distribution. Note that $\sigma = 0$ is not excluded.
Note further that the exponents in the rate conditions (2.6) - (2.8) are precisely what is required to guarantee the absolute summability of the series defining σ^2 (cf.(2.9)), since

$$|E(X_n X_o)| = |E(X_o E(X_n|\mathfrak{F}_o))| \leq \begin{cases} \|X_o\|_2^2 \ \rho(n) \\ 2\|X_o\|_2 \|X_o\|_{2+\delta} \ \phi(n)^{\frac{1+\delta}{2+\delta}} \\ 2(2^{\frac{1+\delta}{2+\delta}} + 1)\|X_o\|^2_{2+\delta} \ \alpha(n)^{\frac{\delta}{2+\delta}} \end{cases}$$

where the first two inequalities follow from Schwartz's inequality and (2.2), respectively (2.3), and the last one from Hölder's inequality and (2.4) with $p=u=\frac{2+\delta}{1+\delta}$ and $q=r=2+\delta$.

Another classical result is the functional CLT which concerns the convergence of the distribution of the process

$$(2.10) \qquad \hat{S}_n(t) = \frac{1}{\sqrt{n}} \sum_{i=1}^{[nt]} X_i + \frac{1}{\sqrt{n}} (nt - [nt])X_{[nt]+1}$$

for $t \in [0,\infty)$.

<u>(2.11) Theorem</u> (Ibragimov [5], Davydov [7]):
Suppose $\{X_i\}_{i\in\mathbb{Z}}$ is a stationary ρ, respectively ϕ, respectively α-mixing sequence. If $E(|X_o|^{2+\delta}) < \infty$, and

$$(2.12) \qquad \sum_{n=o}^{\infty} \rho(n) < \infty \quad , \ \delta \geq 0$$

respectively

$$(2.13) \qquad \sum_{n=o}^{\infty} \phi(n)^{\frac{1+\delta}{2+\delta}} < \infty, \ \delta \geq 0$$

respectively

$$(2.14) \qquad \sum_{n=o}^{\infty} \alpha(n)^{\frac{\delta}{2(2+\delta)}} < \infty \ , \ \delta > 0$$

then

$$(\hat{S}_n(t))_{t\geq o} \Rightarrow \sigma(W(t))_{t\geq o}$$

where $(W(t))_{t\geq o}$ is a standard Wiener process and σ is given by (2.9). □

These two theorems turn out to be immediate consequences of our analysis, which furthermore directly gives conditions precluding $\sigma^2=0$.

III. Martingale Approximation

Let us first recall the CLT for martingale differences:

(3.1) Theorem ([1],[2])

Let $\{m_i\}_{i\in\mathbb{Z}}$ denote a stationary ergodic sequence of square integrable martingale differences adapted to an increasing family of σ-algebras $\{\mathcal{F}_i\}_{i\in\mathbb{Z}}$, i.e. $m_i\in\mathcal{F}_i$, $E(m_i|\mathcal{F}_{i-1}) = 0$ for all i. Then

$$\frac{1}{\sqrt{n}}\sum_{i=1}^{n} m_i \Rightarrow \mathcal{N}(0,\sigma^2) \quad \text{as } n\to\infty$$

where $\sigma^2 = E(m_o^2)$.

Moreover the functional version holds namely

$$(\hat{M}_n(t))_{t\geq 0} \Rightarrow \sigma(W(t))_{t\geq 0}\ , \quad \text{as } n\to\infty$$

where $\hat{M}_n(t)$ is defined like $\hat{S}_n(t)$ (c.f.(2.10)) using m_i instead of X_i. □

Note that, though martingale differences are uncorrelated, i.e. $E(m_i m_j) = 0$ for $i\neq j$, no mixing properties are implied. For example we may take

$$m_i = \xi_i y_i$$

where $\{y_i\}_{i\in\mathbb{Z}}$ is any process (with $E(|y_i|) < \infty$), not necessarily having good mixing properties and $\{\xi_i = +(-)1\}_{i\in\mathbb{Z}}$ is a centered i.i.d. family independent of $\{y_i\}_{i\in\mathbb{Z}}$. Then the m_i's are clearly martingale differences but $|m_i| = |y_i|$; so $\{m_i\}_{i\in\mathbb{Z}}$ need not have good mixing properties.

Now, let $\{X_i\}_{i\in\mathbb{Z}}$ be any stationary ergodic sequence of centered square integrable random variables adapted to an increasing family of σ-algebras $\{\mathcal{F}_i\}_{i\in\mathbb{Z}}$.

Suppose we can find a function $\tilde{g}\in\mathcal{F}_o$ such that

(3.2) $$\tilde{g}_{i+1} - \tilde{g}_i + X_i = m_{i+1}$$

is a sequence of square integrable martingale differences, where $\tilde{g}_i = U^i\tilde{g}$, and U is the operator on functions induced by an automorphism T of $(\Omega,\mathcal{F},\mathbb{P})$, such that $UX_i = X_{i+1}$ for all i and $T^{-1}\mathcal{F}_i = \mathcal{F}_{i+1}$.

Then, by telescoping ($\tilde{g}_0 \equiv \tilde{g}$)

$$(3.3) \qquad \frac{1}{\sqrt{n}}(\tilde{g}_n - \tilde{g}_0) + \frac{1}{\sqrt{n}} S_n = \frac{1}{\sqrt{n}} \sum_{i=1}^{n} m_i \quad .$$

Thus, by Theorem (3.1)

$$\frac{1}{\sqrt{n}} S_n \Rightarrow \mathcal{N}(0,\sigma^2)$$

with $\sigma^2 = E(m_0^2)$ since by stationarity

$$(3.4) \qquad \frac{1}{\sqrt{n}}(\tilde{g}_n - \tilde{g}_0) \to 0$$

in probability.
Clearly it is enough to satisfy (3.2) with i=0, so that (3.2) reduc to

$$E(U\tilde{g} - \tilde{g} + X_0|\mathfrak{F}_0) = 0$$

i.e.

$$(3.5) \qquad (I - P)\tilde{g} = X_0$$

where

$$P = E_{\mathfrak{F}_0} U \qquad (E_{\mathfrak{F}_0} = E(\cdot|\mathfrak{F}_0)) \quad .$$

The formal solution to (3.5) is

$$(3.6) \qquad \tilde{g} = (I - P)^{-1} X_0 = \sum_{n=0}^{\infty} P^n X_0 \quad .$$

Since

$$UE_{\mathfrak{F}_i} = E_{\mathfrak{F}_{i+1}} U$$

it follows that

$$P^n = E_{\mathfrak{F}_0} U^n \quad .$$

Therefore (3.6) may be written as

$$(3.7) \qquad \tilde{g} = \sum_{n=0}^{\infty} E(X_n|\mathfrak{F}_0) \quad .$$

If the series (3.7) is convergent in L_2 then $\tilde{g}$ satisfies (3.5) and the CLT follows.
It also follows then that

$$\sum_{n=0}^{\infty} E(X_n X_0) = E(\tilde{g} X_0)$$

so that by (3.5)

$$\sigma^2 = E(m_1^2) = E((\tilde{g}_1 - \tilde{g}_0 + X_0)^2) = 2\, E(\tilde{g}X_0) - E(X_0^2)$$

$$= \sum_{n=-\infty}^{\infty} E(X_n X_0) \quad .$$

Moreover by stationarity for any $T < \infty$ and any $\varepsilon > 0$

$$\mathbb{P}(\sup_{1<k<[nT]} |\tilde{g}_k - \tilde{g}_0| > \varepsilon \sqrt{n})$$

$$< 2[nT]\, \mathbb{P}(|\tilde{g}| > \frac{\varepsilon}{2} \sqrt{n})$$

$$< \frac{8[nT]}{n\varepsilon^2} \int \chi(|\tilde{g}| > \frac{\varepsilon}{2}\sqrt{n})\tilde{g}^2 \, d\,\mathbb{P} \to 0 \quad \text{as } n \to \infty \quad .$$

Of course, the series (3.7) is convergent in L_2 if

$$(3.8) \qquad \sum_{n=0}^{\infty} \| E(X_n | \mathfrak{F}_0) \|_2 < \infty \quad .$$

Hence Theorem (2.5) for ρ and ϕ-mixing and Theorem (2.11) follow immediately using (2.2) - (2.4). Observe however that Theorem (2.5) for α-mixing, with the exponent $\delta/(2+\delta)$ rather than $\delta/2(2+\delta)$, does not follow from the analysis given so far.

Note that if (3.7) is convergent only in L_p with $p<2$, then (3.2) is satisfied except that m_i need not be in L_2. It will, however, be in L_2 if the formal difference $\tilde{g}_{i+1} - \tilde{g}_i$ is in L_2. In fact it is not required (for the CLT argument) even that $\tilde{g}$ exists as a function - the argument given requires merely that the formal series

$$(3.9) \qquad \tilde{g}_{i+1} - \tilde{g}_i \in L_2$$

and that the formal series

$$(3.10) \qquad \frac{1}{\sqrt{n}} (\tilde{g}_n - \tilde{g}_0) \to 0$$

in probability.

We now make this remark more precise and at the same time give a condition which guarantees (3.9) and (3.10). Note first that formally

$$(3.11) \quad \tilde{g}_1 - \tilde{g}_0 + X_0 = \sum_{n=0}^{\infty} E(X_n | \mathfrak{F}_1) - E(X_n | \mathfrak{F}_0) \equiv m_1 \quad .$$

It follows from the Hilbert space lemma (5.1) with $X = X_0$ and $P_0 = E_{\mathfrak{F}_0}$

that if (3.15) below is satisfied, the series is convergent in L_2, and that

(3.12) $$\frac{1}{\sqrt{n}} \sum_{i=0}^{n} g_i \to 0$$

in L_2, where $g_i = U^{i-1} g_1$ and

(3.13) $$g_1 = m_1 - X_0 \quad .$$

Moreover $\{m_i = U^{i-1} m_1\}_{i \in \mathbb{Z}}$ is manifestly an adapted sequence of martingale differences.

We thus have

<u>(3.14) Theorem:</u>

Suppose that

(3.15) $$\sup_{N \geq M} \sup_{n \geq M} \left| \sum_{k=M}^{N} E(E(X_k | \mathcal{F}_0) X_n) \right| \equiv \varepsilon(M) \to 0 \quad \text{as } M \to \infty \; .$$

Then

$$\frac{1}{\sqrt{n}} S_n \Rightarrow \mathcal{N}(0, \sigma^2)$$

where

$$\sigma^2 = \lim_{n \to \infty} \frac{1}{n} E(S_n^2) \quad . \qquad \square$$

Note that the formula for σ^2 follows from (3.12).

If $E(|X_0|^{2+\delta}) < \infty$, $\delta \geq 0$, it follows from Hölder's inequality that (3.15) is implied by

(3.16) $$\sum_{k=0}^{\infty} \| E(X_k | \mathcal{F}_0) \|_p < \infty$$

provided $p \geq \frac{2+\delta}{1+\delta}$.

Thus, by (2.4),

$$\| E(X_k | \mathcal{F}_0) \|_{\frac{2+\delta}{1+\delta}} \leq \text{const } \alpha(k)^{\frac{\delta}{2+\delta}} \, \| X_0 \|_{2+\delta}$$

and the α-mixing part of Theorem (2.5) follows.

<u>Remark:</u> McLeish has shown [8] that if for some $q < 2$, $\sum_{i=0}^{\infty} \| E(X_i | \mathcal{F}_0) \|_2^q <$ then the measures induced by $(\hat{S}_n(t))_{t \geq 0}$ form a tight family.

Therefore Theorem (2.11) remains valid if the α-mixing exponent is replaced by any number smaller than $\delta/(2+\delta)$. □

Note that it follows from (3.16) that

$$g_1 = \tilde{g}_1 - \tilde{g}_0$$

with $\tilde{g}_0 \in L_p$ $(p \geq \frac{2+\delta}{1+\delta})$.

Therefore we have the

<u>(3.17) Corollary:</u> Suppose $E(|X_0|^{2+\delta}) < \infty$. If (3.16) $(p \geq \frac{2+\delta}{1+\delta})$ is satisfied and $\sigma^2 = 0$, then

(3.18) $$\sup_n \|S_n\|_p < \infty \quad .$$

In particular the distribution of S_n, $n \geq 0$ form a tight family. □

<u>Proof:</u> Since $\sigma^2 = E(m_1^2)$, $\sigma^2 = 0$ implies $m_1 = 0$, and therefore

$$S_n = \tilde{g}_0 - \tilde{g}_n \quad \text{if} \quad \sigma^2 = 0$$

and

$$\|S_n\|_p \leq 2\|\tilde{g}\|_p < \infty \quad . \quad \square$$

Recall that the ρ and ϕ-mixing parts of Theorem (2.5) imply (3.16) for $p = 2$. In this case, if $\sigma^2 = 0$ then $E(S_n^2)$ is bounded [6] . Similarly for the α-mixing part: if $\sigma^2 = 0$ then $E((S_n)^{\frac{2+\delta}{1+\delta}})$ is bounded.

<u>Example:</u> Let $\{X_i\}_{i\in \mathbf{Z}}$ be a finite state space Markov chain which is ergodic and aperiodic.

Then $\sum_{n=0}^{\infty} \|E(X_n|X_0)\|_\infty < \infty$ and hence if $\sigma^2 = 0$ then S_n is bounded uniformly in n and ω . □

Another corollary of Theorem (3.14) is the following

<u>(3.19) Corollary:</u> Let $X_i = X(\xi_i)$, where $\underline{\xi} = \{\xi_i\}_{i\in \mathbf{Z}}$ is a reversible ergodic Markov process.

If

(3.20) $$\sum_{n=0}^{\infty} |E(X_n X_0)| < \infty$$

then

$$\frac{1}{\sqrt{n}} S_n \to \mathcal{N}(0,\sigma^2)$$

with

$$\sigma^2 = \sum_{n=-\infty}^{\infty} E(X_n X_0) \quad . \qquad \square$$

Proof: Let $\mathcal{P} \equiv P(\xi, d\eta)$ be the transition probability. Then by the reversibility of ξ $\mathcal{P}$ is selfadjoint on $L_2(d\mu)$, where μ is the reversible measure. Then, with $\mathcal{F}_0 = \sigma(\xi_i \ , \ i \leqslant 0)$,

$$E(E(X_k|\mathcal{F}_0)X_n) = \int d\mu\, \mathcal{P}^k X\, \mathcal{P}^n X = \int d\mu\, X\, \mathcal{P}^{k+n} X = E(X_0 X_{k+n}) \quad .$$

Thus the condition (3.15) is satisfied if (3.20) holds, and by Theorem (3.14), with U induced by the shift on $\{\xi_i\}_{i \in \mathbb{Z}}$, the result follows. $\square$

Remark: Kipnis and Varadhan [9] have shown that in this case (3.20) is also sufficient for the invariance principle.

Remark: The reader should compare (3.16) with the (ρ,ϕ,α)-mixing conditions in (2.12) - (2.14). Note that

$$\| E(X_n|\mathcal{F}_0)\|_p = \sup_{\substack{f \in \mathcal{F}_0 \\ \|f\|_q \leqslant 1}} E(f\, X_n) \ , \text{ where } p^{-1} + q^{-1} = 1$$

does not invoke any uniformity in $h (h \in \mathcal{F}^n)$ at all; in fact, for each n only one function h is involved namely $h = X_n$.

IV A more general result

So far we considered only the case where the sequences $\{X_i\}_{i \in \mathbb{Z}}$ are adapted to the family of σ-algebras $\{\mathcal{F}_i\}_{i \in \mathbb{Z}}$. There are situations, however, where X_0 is a function of some other process $\{\xi_i\}_{i \in \mathbb{Z}}$, which is "good mixing" (w.r.t. the family of σ-algebras $\{\mathcal{F}_i^\xi\}_{i \in \mathbb{Z}}$ generated by the ξ_i's i.e. $\mathcal{F}_i \equiv \mathcal{F}_i^\xi = \sigma(\xi_j, \ j \leqslant i)$). Then $X_0 = X(\{\xi_i\}_{i \in \mathbb{Z}})$ will in general not be $\mathcal{F}_0^\xi$ measurable. One might think for example of a dynamical system, given by an ergodic automorphism T on the measure space $(\Omega, \mathcal{F}, \mu)$, which is isomorphic to a symbolic dynamics with good mixing properties. This case is easily handled.

We write

$$X_o = Z_o + Z_o^+$$

where

$$Z_o = E(X_o|\mathcal{F}_o)$$

and apply the Hilbert space lemma (5.1), first with $X = Z_o$ and with $P_o = E_{\mathcal{F}_o}$ to obtain that if (4.5) below holds, then

(4.1) $$g_1 = m_1 - Z_o$$

where $m_1 \in \mathcal{F}_1$, and $E(m_1|\mathcal{F}_o) = 0$ and g_i satisfies (3.12).

We then apply the Hilbert space lemma (5.1) with $X = U^{-1}Z_o^+$, $P_o = E_{\mathcal{F}_o^+} \equiv I - E_{\mathcal{F}_o}$ and U^{-1} now playing the role of U. We then obtain that if (4.6) below is satisfied, then

(4.2) $$g = m - U^{-1}Z_o^+$$

where $m \in \mathcal{F}_o$ and $E(m|\mathcal{F}_{-1}) = 0$ and $\frac{1}{\sqrt{n}} \sum_{k=o}^{n-1} U^{-k} g \to 0$ in L_2.

Applying U to (4.2) yields ($h_1 = Ug$)

(4.3) $$h_1 = n_1 - Z_o^+$$

where $n_1 \in \mathcal{F}_1$, $E(n_1|\mathcal{F}_o) = 0$ and

$$\frac{1}{\sqrt{n}} \sum_{k=o}^{n-1} h_i \to 0 \quad \text{in } L_2 .$$

Thus, by adding (4.1) and (4.3), we obtain a decomposition of X_o of the form (3.13). We therefore have the

<u>(4.4) Theorem:</u>

Suppose that

(4.5) $$\sup_{N \geq M} \sup_{n \geq M} \left| \sum_{k=M}^{N} E(E(X_k|\mathcal{F}_o)X_n) \right| \equiv \varepsilon(M) \to 0 \quad \text{as } M \to \infty$$

and that

(4.6) $$\sup_{N \geq M} \sup_{n \geq M} \left| \sum_{k=M}^{N} E((X_{-k} - E(X_{-k}|\mathcal{F}_o))X_{-n}) \right| \equiv \tilde{\varepsilon}(M) \to 0 \quad \text{as } M \to \infty .$$

Then

$$\frac{1}{\sqrt{n}} S_n \Rightarrow \mathcal{N}(0,\sigma^2) \quad ,$$

where $\sigma^2 = \lim\limits_{n\to\infty} \frac{E(S_n^2)}{n}$. □

If $E(|X_0|^{2+\delta}) < \infty$ we obtain by Hölder's inequality that(4.5) and (4.6) are implied by $(p \geq (2+\delta)/(1+\delta))$

$$(4.7) \qquad \sum_{n=0}^{\infty} \| E(X_n|\mathcal{F}_0)\|_p + \sum_{n=-\infty}^{-1} \| X_n - E(X_n|\mathcal{F}_0)\|_p < \infty$$

a condition first obtained by Gordin.

<u>Remark:</u> The previous discussion depends crucially on the choice of $\mathcal{F}_0$; for example the choice $\mathcal{F}_0 = \mathcal{F}$, the full σ-algebra is useless.

The question thus naturally arises as to whether the CLT can be satisfied in a situation where there is no decomposition of the form (3.13) regardless of the choice of $\mathcal{F}_0$. To make the question sharper consider a dynamical system having zero Kolmogorov-Sinai-Entropy. This implies that , whenever $\ldots\mathcal{F}_{-1} \subset \mathcal{F}_0 \subset \mathcal{F}_1 \ldots (\mathcal{F}_n = T^{-n}\mathcal{F}_0)$,then in fact $\mathcal{F}_n = \mathcal{F}_0$ for all n. Thus

$$0 = E(m_i|\mathcal{F}_{i-1}) = E(m_i|\mathcal{F}_i) = m_i$$

and no argument based on a martingale difference approximation as described here can lead to a CLT with $\sigma^2 > 0$. The question then arises as to whether there exists any dynamical system with entropy zero having a function X with $E(X)=0$,$E(X^2)< \infty$ satisfying a CLT with $\sigma^2 > 0$.

<u>V. A Hilbert space lemma</u>

<u>(5.1) Lemma:</u>

Let $\mathcal{H}$ be a Hilbert space with inner product (,); P_0 an orthogonal projection on $\mathcal{H}$ and U a unitary satisfying

$$U P_0 U^{-1} \equiv P_1 \geq P_0$$

i.e.,$P_1P_0 = P_0P_1 = P_0$. Let $X \in \mathcal{H}$ and suppose that $P_1X = X$ and that

$$\sup_{N\geq M} \sup_{n\geq M} \left| \sum_{k=M}^{N} (P_0X_k, X_n)\right| \equiv \varepsilon(M) \to 0 \quad \text{as } M \to \infty$$

where $X_k = U^kX$.

Then

(5.2) $\sum_{n=o}^{\infty} P_{o,1} X_n$ is convergent in $\mathcal{H}$

where $P_{o,1} = P_1 - P_o$.

Let

$$m = \sum_{n=o}^{\infty} P_{o,1} X_n$$

and

$$g = m - X \quad .$$

Then

(5.3) $\frac{1}{\sqrt{n}} \sum_{i=o}^{n-1} g_i \to 0$ in $\mathcal{H}$

where $g_i = U^{i-1} g$. □

<u>Proof:</u> To establish (5.2) we show that $\sum_{n=o}^{\infty} P_{o,1} X_n$ is a Cauchy series in $\mathcal{H}$, i.e. that for $\varepsilon > o$ there exists M such that for all $N \geqslant M$

$$\| \sum_{n=M}^{N} P_{o,1} X_n \| < \varepsilon \quad .$$

Let $P_j \equiv U^j P_o U^{-j}$ and $P_{j,k} \equiv P_k - P_j$ for $j < k$. Then

$$\| \sum_{n=M}^{N} P_{o,1} X_n \|^2$$

$$= \sum_{n=M}^{N} (P_{o,1}X_n, P_{o,1}X_n) + 2 \sum_{m>n=M}^{N} (P_{o,1}X_n, P_{o,1}X_m)$$

$$= \sum_{n=M}^{N} (P_{-n,-n+1} X, P_{-n,-n+1}X) + 2 \sum_{m=n+1}^{N} \sum_{n=M}^{N-1} (P_{-n,-n+1}X, P_{-n,-n+1}X_{m-n})$$

$$= \sum_{n=M}^{N} (P_{-n,-n+1}X, P_{-n,-n+1}X) + 2 \sum_{k=1}^{N-M} \sum_{n=M}^{N-k} (P_{-n,-n+1}X, P_{-n,-n+1}X_k)$$

$$= (P_{-N,-M+1}X, P_{-N,-M+1}X) + 2 \sum_{k=1}^{N-M} (P_{-(N-k),-M+1}X, P_{-(N-k),-M+1}X_k)$$

$$= (P_{-M+1}X , P_{-M+1}X) - (P_{-N}X,P_{-N}X)$$

$$+ 2 \sum_{k=1}^{N-M} ((P_{-M+1}X,P_{-M+1}X_k) - (P_{-(N-k)}X,P_{-(N-k)}X_k))$$

$$= (P_0 X_{M-1} , P_0 X_{M-1}) - (P_0 X_N,P_0 X_N)$$

$$+ 2 \sum_{k=M}^{N-1} ((P_0 X_{M-1} , P_0 X_k) - (P_0 X_k,P_0 X_N))$$

$$= 2 \sum_{k=M-1}^{N} ((P_0 X_k,X_{M-1}) - (P_0 X_k,X_N)) + (P_0 X_N,X_N) - (P_0 X_{M-1},X_{M-1})$$

$$\leqslant 6\, \varepsilon(M-1)$$

This proves (5.2).

For (5.3) we note that

$$(5.4) \qquad \sum_{i=1}^{k} g_i = \sum_{i=1}^{k} g_i^{(k-i+1)} - \sum_{i=0}^{k-1} P_0 X_i$$

where

$$g_i^{(m)} = U^{i-1} g_1^{(m)}$$

and

$$g_1^{(m)} = \sum_{j=m}^{\infty} P_{0,1} X_j$$

and where we have used that $P_1 X = X$.

Now

$$\frac{1}{k} \| \sum_{i=1}^{k} g_i^{(k-i+1)} \|^2 = \frac{1}{k} \sum_{m=1}^{k} \| g_1^{(m)} \|^2 \to 0 \quad \text{as } k \to \infty$$

by (5.2).

Moreover

$$\| \sum_{i=0}^{k-1} P_0 X_i \| \leqslant \| \sum_{i=0}^{M} P_0 X_i \| + \| \sum_{i=M+1}^{k-1} P_0 X_i \|$$

and

$$\| \sum_{i=M+1}^{k-1} P_0 X_i \|^2 \leqslant k\varepsilon(M) \quad .$$

(5.3) follows now easily. □

Acknowledgements: We thank the IHES for their hospitality while this was being written.

References:

[1] Billingsley,P.: Convergence of Probability Measures, John Wiley and Sons, 1968.

[2] Ibragimov, I.A.: A Central Limit Theorem for a class of dependent random variables, Theor.Prob.Appl.8, 83-89,1963.

[3] Gordin,M.I.: On the Central Limit Theorem for stationary processes. Soviet Math.Dokl.,10, 1174-1176, 1969.

[4] Hall,P., Heyde,C.: Martingale limit theory and its application. Academic Press, New York, 1980.

[5] Ibragimov,I.A.: A Note on the CLT for dependent random variables. Theor.Prob.Appl.20, 135-140, 1975.

[6] Ibragimov,Linnik: Independent and stationary sequences of random variables. Wolters Noordhoff Publishing, Groningen 1969.

[7] Davydov,Y.A.: The invariance principle for stationary processes. Theor.Prob.Appl.15, 487-498, 1970.

[8] McLeish,D.L.: Invariance principles for dependent variables. Z.Wahrsch.verw.Gebiete, 32, 165-178, 1975.

[9] Kipnis,C.,Varadhan,S.R.S.: Central Limit Theorem for additive functionals of reversible Markov processes and application to simple exclusions: to appear in Comm.Math.Phys.,1985.

TIME REVERSAL ON WIENER SPACE

H. Föllmer
Mathematikdepartement
ETH Zentrum, CH-8092 Zürich

1. Introduction

Let P be a probability measure on C[0,1] which has finite relative entropy

(1.1) $$H(P|P^*) = E[\log \frac{dP}{dP^*}] < \infty$$

with respect to Wiener measure P*. It follows from the theory of the Girsanov transformation that the coordinate process (X_t) satisfies the stochastic differential equation

(1.2) $$dX_t = dW_t + b_t dt \qquad (0 \leq t \leq 1)$$

where (W_t) is a Wiener process under P, and where (b_t) is an adapted process with finite energy

(1.3) $$E[\int_0^1 b_t^2 dt] < \infty .$$

It is also clear that the distribution of X_t under P is absolutely continuous with some density function ρ_t for any $t > 0$.

Now consider the time reversal $\hat{P} = P \circ R$ where R denotes the pathwise time reversal on C[0,1]. Under $\hat{P}$, the coordinate process is again governed by a stochastic differential equation of the form (1.2) with some adapted process $(\hat{b}_t)$. The first purpose of this paper is to give a rigorous derivation of the duality equation

(1.4) $$E[\hat{b}_{1-t} \circ R + b_t | X_t = x] = (\log \rho_t(x))'$$

for almost all t and for ρ_t-almost all x. In particular we show that the density function ρ_t is almost everywhere differentiable; no regularity assumption on the drift (b_t) is needed, only the finite energy condition (1.3).

In the Markovian case, (1.4) reduces to the well known equation

(1.5) $$\hat{b}_{1-t}(x) + b_t(x) = (\log \rho_t(x))'$$

which expresses the dual drift in terms of the successive densities of the process; cf.,e.g., [11,12, 1, 4, 7, 14]. But in the general non-Markovian case, (1.4) does not yet specify the dual drift, only its projection on the present state. For a full description, we have to replace the integration by parts on R^1 which is implicit in (1.4) resp. (1.5), by an integration by parts on Wiener space, i.e., by a Malliavin calculus argument. This leads to a path space formula for $(\hat{b}_t)$; cf. (4.8).

Our method is based on the fact that the finite entropy condition (1.1) is invariant under time reversal. The argument extends without change to the finite-dimensional case. For infinite-dimensional diffusion processes, one needs a suitably localized version of the entropy technique; this has been worked out in a joint paper with A.Wakolbinger [13]. The results of the present paper were announced in [6]; here we give detailed proofs for section 3 of [6].

2. Drifts as forward derivatives

Let us denote by (X_t) the coordinate process and by $(\mathcal{F}_t)$ the canonical filtration on $C[0,1]$, and let P be a probability measure on $C[0,1]$ with initial distribution $\mu = P \circ X_0^{-1}$. We denote by P_x^* the Wiener measure with starting point x and assume that P is absoultely continuous with respect to $P^* \equiv \int \mu(dx) P_x^*$. This is equivalent to the existence of an adapted process (b_t) with $\int_0^1 b_t^2 dt < \infty$ P-a.s. such that

(2.1) $$W_t^b \equiv X_t - X_0 - \int_0^t b_s ds \qquad (0 \le t \le 1)$$

is a Wiener process under P ; cf. [5],[9]. The process (b_t) will be called the drift of P .

(2.2) Remark. Let (G_t) be a continuous version of the martingale

$$G_t = \frac{dP}{dP^*}\Big|_{\mathcal{F}_t} \qquad (0 \le t \le 1)$$

and note that $\inf G_t > 0$ P-a.s.. Itô's representation of (G_t) as a stochastic integral

$$G_t = \int_0^t H_s dX_s \quad , \qquad \int_0^1 H_s^2 ds < \infty \qquad P^*\text{-a.s.}$$

and the general properties of the Girsanov transformation [5,8] imply that (b_t) is given by

(2.3) $$b_t = H_t G_t^{-1} \qquad dP^* \times dt\text{-a.s.}$$

Thus, the drift (b_t) can be computed in terms of (G_t) by means of a pathwise differentiation of the covariance process $\langle G,X\rangle_t = \int_0^t H_s ds$. Conversely, a P-almost sure application of Itô's formula to the paths of $\log G_t$ shows that (G_t) can be expressed in terms of (b_t) by the Girsanov formula

(2.4) $$G_t = \exp\left[\int_0^t b_t dX_t - \frac{1}{2}\int_0^t b_t^2 dt\right] = \exp\left[\int_0^t b_t dW_t^b + \frac{1}{2}\int_0^t b_t^2 dt\right].$$

For our purpose it will be convenient to use, instead of (2.3), a direct computation of the drift (b_t) as a stochastic forward derivative of the process (X_t) in the sense of Nelson [12]. The following proposition was proved jointly with A.Wakolbinger; cf. also Grigelionis [8] for a closely related result .

(2.5) Proposition. Suppose that

(2.6) $$E\left[\int_0^1 |b_t|^p dt\right] < \infty$$

for some $p \geq 1$. Then we have, for almost all $t \in (0,1)$,

(2.7) $$b_t = \lim_{h\downarrow 0} \frac{1}{h} E[X_{t+h}-X_t \mid \mathcal{F}_t] = \lim_{\alpha,\beta\downarrow 0} \frac{1}{\alpha+\beta}\int_{t-\alpha}^{t+\beta} b_s ds \qquad \text{in } L^p(P);$$

the second limit also exists P-a.s.

Proof. Put

$$F_{\alpha,\beta}(t,\omega) = \frac{1}{\alpha+\beta}\int_{t-\alpha}^{t+\beta} b_s(\omega) ds .$$

By Fubini, we have

(2.8) $$\lim_{\alpha,\beta\downarrow 0} F_{\alpha,\beta}(t,.) = b_t \qquad \text{P-a.s.} ,$$

for almost all $t \in (0,1)$.

1) Let us first assume $b_t \geq 0$ and $p = 1$. For almost all $t \in (0,1)$ we have

$$\lim_{\alpha,\beta\downarrow 0} E[F_{\alpha,\beta}(t,.)] = \lim_{\alpha,\beta\downarrow 0} \frac{1}{\alpha+\beta}\int_{t-\alpha}^{t+\beta} E[b_s] ds = E[b_t]$$

in addition to (2.8). But this implies, because of our assumption $b_s \geq 0$, that we have also $L^1(P)$-convergence in (2.8):, cf.[10]II.T21.

2) The argument in 1) shows that, for almost all $t \in (0,1)$,

$$\lim \frac{1}{\alpha+\beta} \int_{t-\alpha}^{t+\beta} |b_s|^p ds = |b_t|^p \qquad \text{in } L^1(P).$$

Thus,

$$|F_{\alpha,\beta}(t,.) - b_t|^p \leq 2^p |F_{\alpha,\beta}(t,.)|^p + |b_t|^p$$
$$\leq 2^p \frac{1}{\alpha+\beta} \int_{t-\alpha}^{t+\beta} |b_s|^p ds + |b_t|^p$$

is uniformly integrable, and this implies that the convergence in (2.8) holds not only P-a.s. but also in $L^p(P)$. The first equation in (2.7) now follows by contraction:

$$E[|\frac{1}{h} E[X_{t+h} - X_t | \mathcal{F}_t] - b_t|^p] \leq E[|\frac{1}{h} \int_t^{t+h} b_s ds - b_t|^p] .$$

We are going to apply (2.7) in the case $p = 2$ where (2.6) is just the finite energy condition (1.3). In view of time reversal, it will be useful to restate this condition in terms of entropy.

(2.9) Definition. Let us say that P has finite entropy with respect to Wiener measure if

(2.10) $$H(P|P^*) \equiv \int \log \frac{dP}{dP^*} \, dP < \infty .$$

(2.11) Proposition. P has finite entropy with respect to Wiener measure if and only if its drift (b_t) satisfies the finite energy condition (1.3), and in that case we have

(2.12) $$H(P|P^*) = \frac{1}{2} E[\int_0^1 b_t^2 dt]$$

Proof: The Girsanov formula (2.4 implies

$$H(P|P^*) = E[\int_0^1 b_t dW_t^b] + \frac{1}{2} E[\int_0^1 b_t^2 dt] ,$$

and the first term on the right vanishes if condition (1.3) is satisfied. The converse is contained in [6] lemma (2.6).

3. Time reversal and the duality equation

We denote by $\hat{P} = P \circ R$ the image of P under the pathwise time reversal R on $C[0,1]$ defined by $X_t(R(\omega)) = X_{1-t}(\omega)$.

(3.1) Lemma. If P has finite entropy with respect to Wiener measure then its time reversal $\hat{P}$ has finite entropy up to any time $t<1$, i.e., there exists an adapted process $(\hat{b}_t)$ such that

(3.2) $$\hat{W}_t \equiv X_t - X_0 - \int_0^t \hat{b}_s ds \qquad (0 \le t \le 1)$$

is a Wiener process under $\hat{P}$, and

(3.3) $$\hat{E}\left[\int_0^t \hat{b}_s^2 ds\right] < \infty \qquad \text{for any } t<1 .$$

Proof: With respect to the time reversal $P^* \circ R$ of P^*, the coordinate process is a Brownian motion conditioned to have distribution μ at time $t = 1$. This implies that

(3.4) $$W_t^1 \equiv X_t - X_0 - \int_0^t c_s ds, \qquad c_s \equiv \frac{X_1 - X_s}{1-s}$$

is a Wiener process with respect to $P^* \circ R$ and the filtration $\mathcal{A}_t = \mathcal{F}_t \vee \sigma(X_1)$. Since relative entropy does not increase under a measurable transformation, we obtain $H(\hat{P}|P^* \circ R) = H(P|P^*)$ and in particular

(3.5) $$H(\hat{P}|P^* \circ R) < \infty .$$

But (3.5) implies that there is a process (a_t) adapted to $(\mathcal{A}_t)$ with

$$\hat{E}\left[\int_0^1 a_t^2 dt\right] < \infty$$

such that

(3.6) $$W_t^2 \equiv W_t^1 - \int_0^t a_s ds \qquad (0 \le t \le 1)$$

is a Wiener process with respect to $\hat{P}$ and $(\mathcal{A}_t)$; cf. [6] lemma (2.6). On the other hand,

$$\sup_s \hat{E}[(X_1 - X_s)^2] = \sup_s E[(X_s - X_0)^2]$$

due to (2.1) and (2.12), and so we get

$$(3.7) \qquad E\left[\int_0^t (a_s + c_s)^2 \, ds\right] < \infty \qquad \text{for any } t < 1 .$$

Now let $(\hat{b}_t)$ denote an optional version of the process $\hat{E}[a_t + c_t \mid \mathcal{F}_t]$ $(0 \leq t \leq 1)$ and put

$$(3.8) \qquad \hat{W}_t \equiv X_t - X_0 - \int_0^t \hat{b}_s ds \qquad (0 \leq t \leq 1).$$

Then $(\hat{b}_t)$ satisfies (3.3) due to (3.7), and $(\hat{W}_t)$ is a Wiener process with respect to $\hat{P}$ and $(\mathcal{F}_t)$, due to (3.4) and (3.6).

From now on we assume that P has finite entropy with respect to Wiener measure. (3.2) shows that under $\hat{P}$ the coordinate process satisfies a stochastic differential equation of the form (1.2) with some dual drift $(\hat{b}_t)$, and by (3.3) and (2.5) we can compute $(\hat{b}_t)$ as a stochastic forward derivative under $\hat{P}$ or, equivalently, as a stochastic backward derivative under P :

$$(3.9) \qquad \begin{aligned} \hat{b}_t &= \lim_{h \downarrow 0} \frac{1}{h} \hat{E}[X_{t+h} - X_t \mid \mathcal{F}_t] \\ &= -\lim_{h \downarrow 0} \frac{1}{h} E[X_{1-t} - X_{1-t-h} \mid \hat{\mathcal{F}}_{1-t}] \circ R \qquad \text{in } L^2(P) \end{aligned}$$

for almost all $t \in (0,1)$, where $\hat{\mathcal{F}}_{1-t} \equiv R^{-1}(\mathcal{F}_t)$ denotes the σ-algebra of events observable from time t on. Since P is absolutely continuous with respect to P^*, the distribution of X_t under P is absolutely continuous for each $t > 0$, and we denote by $\rho_t(x)$ its density function.

(3.10) <u>Theorem</u>. For almost all $t \in (0,1)$, ρ_t is an absolutely continuous function, and its derivative ρ'_t satisfies the <u>duality equation</u>

(3.11) $$\rho_t'(x) = \rho_t(x) \quad E[\hat{b}_{1-t}\circ R + b_t | X_t = x]$$

for almost all x. For any $t > 0$ the integrability condition

(3.12) $$\int_t^1 \int_{-\infty}^{\infty} \frac{(\rho_s'(x))^2}{\rho_s(x)} dxds = E[\int_t^1 (\frac{\rho_s'}{\rho_s})^2 (X_s) ds] < \infty$$

is satisfied.

Proof. 1) Let t_0 be a point in (0.1) such that (2.7) holds for $(\hat{b}_s)$ and $t = 1-t_0$, and for (b_s) and $t = t_0$. Then we have, for any bounded $\hat{\mathfrak{F}}_{t_0}$-measurable function F on $C[0,1]$,

(3.13) $$\begin{aligned} E[(\hat{b}_{1-t_0}\circ R)\, F] &= \hat{E}[\hat{b}_{1-t_0}\,(F\circ R)] \\ &= \lim \frac{1}{h}\hat{E}[(X_{1-t_0+h} - X_{1-t_0})(F\circ R)] \\ &= -\lim \frac{1}{h} E[(X_{t_0} - X_{t_0-h})\, F] \,. \end{aligned}$$

2) For $F = f(X_{t_0})$ with some smooth function f with compact support we have

$$\begin{aligned} E[(X_{t_0} - X_{t_0-h})\, f(X_{t_0})] &= E[(\int_{t_0-h}^{t_0} b_s ds)\, f(X_{t_0-h})] \\ &\quad + E[(X_{t_0} - X_{t_0-h})(f(X_{t_0}) - f(X_{t_0-h}))] \end{aligned}$$

and

$$f(X_{t_0}) - f(X_{t_0-h}) = \int_{t_0-h}^{t_0} f'(X_s) dX_s + \frac{1}{2}\int_{t_0-h}^{t_0} f''(X_s) ds$$

P^* a.s. hence P-a.s.. By (3.13) we get, using some straightforward estimates based on (2.7),

(3.14) $$E[(\hat{b}_{1-t_0}\circ R)\, f(X_{t_0})] = -E[b_{t_0} f(X_{t_0})] + E[f'(X_{t_0})] \,.$$

Since

$$E[f'(X_{t_0})] = \int f'(x)\, \rho_{t_0}(x) dx = -\langle \rho_{t_0}', f\rangle \,,$$

(3.14) shows that the distributional derivative of ϱ_{t_0} is in fact given by a function, namely by the right side of (3.11). The integrability condition (3.12) now follows from (3.3) and (2.12).

(3.15) Remark. Suppose that (X_t) is a Markov process under P. Then the drift (b_t) is of the form $b_t(\omega) = b_t(X_t(\omega))$ with some measurable function $b_t(x)$. But the Markov property is preserved under time reversal, and so the dual drift is given by some measurable function $\hat{b}_t(x)$, and (3.11) reduces to the classical duality equation (1.5).

4. A path space formula for the dual drift

In the general non-Markovian case, the dual drift is not yet determined by the duality equation (3.11). For a complete description, we have to replace the functionals $F = f(X_{t_0})$ used in the proof of (3.11) by smooth functionals on $C[0,1]$ which depend on the full σ-algebra $\hat{\mathcal{F}}_{t_0}$. In particular, the integration by parts on R^1 which was used in (3.14) has to be replaced by an integration by parts on Wiener space. Let us first recall the basic formula.

For a bounded predictable process (u_t) put $U_t = \int_0^t u_s ds$ and $X_t^{\varepsilon,U} = X_t + \varepsilon U_t \quad (0 \leq t \leq 1)$.

(4.1) Definition. A function $F \in L^2(P^*)$ is called L²-differentiable if there is a measurable process (φ_t) such that, for any bounded predictable process (u_t),

$$(4.2) \qquad DF(.,U) = \lim_{\varepsilon \downarrow 0} \frac{F(X^{\varepsilon,U}) - F(X)}{\varepsilon} = \int_0 u_s \varphi_s ds \qquad \text{in } L^2(P^*).$$

If F is Fréchet-differentiable on $C[0,1]$ with bounded derivative $DF(\omega,dt)$, viewed as a measure on $[0,1]$, then (4.2) holds with $\varphi_s(\omega) = DF(\omega,[s,1])$. L^2-differentiability is enough to use Bismut's proof [3] of the basic formula

$$(4.3) \qquad E^*\left[\left(\int_0^1 u_s dX_s\right) F\right] = E^*[DF(.,U)] = E^*\left[\int_0^1 u_s \varphi_s ds\right] \quad .$$

If F is Fréchet-differentiable with bounded derivative $DF(.,dt)$ and G is L^2-differentiable then (4.3), applied to FG, leads to the following <u>integration by parts</u>:

$$E^*\Big[FG \int_0^1 u_t dX_t\Big] = E^*\big[DF(.,U)G\big] + E^*\big[FDG(.,U)\big] \tag{4.4}$$

(4.5) <u>Remark</u>. If we apply (4.4) to the Girsanov density $G_1 = \frac{dP}{dP^*}$ then we get

$$E\Big[F \int_0^1 u_t dX_t\Big] = E\big[DF(.,U)\big] + E\big[FG_1^{-1}DG_1(.,U)\big]. \tag{4.5}$$

If the drift (b_t) is a bounded smooth function on $C[0,1]\times[0,1]$ with bounded Fréchet derivatives $Db_t(.,ds)$ then G_1 is indeed L^2-differentiable and satisfies

$$DG_1(.,U) = \int_0^1 u_t \gamma_t dt$$

with

$$\gamma_t G_1^{-1} = b_t + \int_t^1 \beta_{t,s} dW_s^b \tag{4.6}$$

and $\beta_{t,s} = Db_s(.,[t,1])$. This follows, e.g., from theorem (A.10) in Bichteler-Jacod [2]. Instead of giving the details we shall simply assume that G_1 is L^2-differentiable and satisfies (4.6). In fact, a more careful analysis shows that, for the purpose of time reversal, it is not really necessary to impose regularity conditions on the drift: as in (3.10), the necessary degree of smoothness is "built in". This point will be discussed elsewhere.

(4.7) <u>Theorem</u>. Suppose that $G_1 = \frac{dP}{dP^*}$ is L^2-differentiable and satisfies (4.6). Then the dual drift is given, for almost all $t \in (0,1)$, by

$$\hat{b}_{1-t} = -E\big[b_t + a_t \,\big|\, \mathcal{F}_t\big] \circ R \tag{4.8}$$

where

$$a_t = \frac{1}{t}\Big(W_t^b - \int_0^t \int_0^1 \beta_{t,s} dW_s^b \, dt\Big) + \int_t^1 \beta_{t,s} dW_s^b \tag{4.9}$$

Proof. We fix t_0 as in the proof of (3.10) so that

$$E[(\hat{b}_{1-t_0}\circ R)F] = -\lim_h \frac{1}{h} E[(X_{t_0}-X_{t_0-h})F] \tag{4.10}$$

for any bounded $\hat{\mathfrak{F}}_{t_0}$-measurable function F on $C[0,1]$. Now we assume that F is Fréchet-differentiable with bounded derivative $DF(.,dt)$; since F is $\hat{\mathfrak{F}}_{t_0}$-differentiable, the measures $DF(.,dt)$ are concentrated on $[t_0,1]$. Applying (4.5) with $u_t = I_{[t_0-h,t_0]}(t)$ we get

$$E[(X_{t_0}-X_{t_0-h})F] = h\,E[DF(.,[t_0,1])] + E[FG_1^{-1}\int_{t_0-h}^{t_0}\gamma_t dt]. \tag{4.11}$$

This together with (4.10) implies

$$-E[(\hat{b}_{1-t}\circ R)F] = E[DF(.,[t,1])] + E[FG_1^{-1}\gamma_t] \tag{4.12}$$

for almost all $t\in(0,1)$. Using again (4.11) with $h = t_0$ we obtain, due to (4.6),

$$\begin{aligned} E[DF(.,[t,1])] &= \frac{1}{t}E[X_tF] - \frac{1}{t}E[F\int_0^t G_1^{-1}\gamma_s ds] \\ &= E[F\frac{1}{t}(W_t^b - \int_0^t\int_t^1 \beta_{t,s}dW_s^b\,dt)] \end{aligned}$$

Thus, we can write (4.12) as

$$E[(\hat{b}_{1-t}\circ R + b_t)F] = -E[a_t\,F],$$

and this implies (4.8).

(4.13) Remark. In the Markovian case with a smooth drift function $b_t(x)$ we have $\beta_{t,s} = b_s'(X_s)I_{[t,1]}(s)$, and (4.8) reduces to

$$(\log \varrho_t(x))' = -\frac{1}{t}E[\int_0^t (1-sb_s'(X_s))dW_s^b \mid X_t=x]. \tag{4.14}$$

References

[1] B.D.O. ANDERSON: Reverse time diffusion equation models. Stoch. Processes and Appl. 12, 313-326 (1982)

[2] K.BICHTELER, J.JACOD: Calcul de Malliavin pour les diffusions avec sauts. Sém.Probabilités XVII, Lecture Notes Math.986, Springer (1983)

[3] J.M.BISMUT: Martingales, the Malliavin Calculus and Hypoellipticity under general Hörmander's conditions. Z. Wahrscheinlichkeitstheorie verw. Geb. 56,469-505 (1981)

[4] E.A.CARLEN: Conservative Diffusions: A constructive approach to Nelson's stochastic mechanics. Thesis, Princeton (1984)

[5] C.DELLACHERIE, P.A.MEYER: Probabilités et Potentiel. Théorie des Martingales. Hermann, Paris (1980)

[6] H.FÖLLMER: An entropy approach to the time reversal of diffusion processes. In: Proc.4th IFIP Workschop on Stochastic Differential Equations, M.Métivier and E.Pardoux Eds., Lecture Notes in Control and Information Sciences, Springer (1985)

[7] U.HAUSSMANN, E.PARDOUX: Time reversal of diffusion processes.In: Proc.4th IFIP Workshop on Stochastic Differential Equations, M.Métivier and E.Pardoux Eds., Lecture Notes in Control and Information Sciences, Springer (1985)

[8] B.GRIGELIONIS: On the structure of Föllmer's measure and some properties of potentials. Lietuvos Matematikos Rinkinys XVI Nr.4,99-103 (1976)

[9] R.S.LIPTSER, A.N.SHIRYAEV: Statistics of Random Processes I. Springer (1977)

[10] P.A.MEYER: Probability and Potentials. Waltham: Blaisdell (1966)

[11] M.NAGASAWA: Time reversion of Markov processes. Nagaya Math.Journal 24,117-204 (1964)

[12] E.NELSON: Dynamical theories of Brownian motion. Princeton UP (1967)

[13] A.WAKOLBINGER, H.FÖLLMER: Time reversal of infinite-dimensional diffusions. To appear.

[14] W.ZHENG, P.A.MEYER: Quelques resultats de mécanique stochastique. Sém. Probabilités XVIII, Springer Lecture Notes in Math. 1059, 223-244 (1984)

LATTICE GAUGE THEORY; HEURISTICS AND CONVERGENCE

Leonard Gross*
Department of Mathematics
Cornell University
Ithaca, New York 14853

§1. Introduction. The machinery associated with the Feynman-Kac formula for a Markov process can be used to convert some problems of operator theory arising in certain quantum systems into problems of measure theory on the path space of the process. If the quantum system is a quantum field theory then the state space is infinite dimensional. For example if physical space is of dimension n then in the scalar field model it is natural to take the state space to be the space $\mathcal{S}'(R^n)$ of real tempered distributions on R^n. In practice the path space may be regarded as a subset of $\mathcal{S}'(R^{n+1})$ for this quantum field. The problem of existence of quantum fields then translates in this case into the problem of existence of certain heuristically defined measures on $\mathcal{S}'(R^{n+1})$. In this note an informal description of some of the desired measures on path space corresponding to quantized Yang-Mills fields will be given. K. Wilson's lattice construction method for these measures will be described along with a theorem asserting the convergence of the lattice measures to the desired limit in one simple case. This exposition is aimed at probabilists. For further details regarding the link between the measure theory discussed here and quantum fields see [6] or [13].

§2. Heuristic description of path space measures. In order to understand, at least at some intuitive level, the meaning of the heuristic expressions defining the Yang-Mills measures it is useful to understand first the precise meaning of similar but simpler expressions. We begin with the familiar case of Wiener measure. Write W for the space of real valued continuous functions, ϕ, on $[0,1]$ such that $\phi(0) = 0$. If f is a bounded continuous (in sup norm) function on W then its integral with respect to Wiener measure, w, may be defined,

*This research was supported in part by N.S.F. Grant MCS 81-02147, in part by the Institute for Mathematics and its Applications at the University of Minnesota and in part by the Institute for Advanced Study in Princeton, NJ, USA.

as is well known, by the equation

(2.1) $\int_W f(\phi)dw(\phi)$

$$= \lim Z_k^{-1} \int_{R^k} f(\phi^k) \exp\left[-\frac{1}{2}\sum_{j=1}^k \left(\frac{x_j - x_{j-1}}{t_j - t_{j-1}}\right)^2 (t_j - t_{j-1})\right] dx_1 \ldots d_{x_k}$$

where $0 = t_0 < t_1 < \ldots < t_k = 1$ is a partition of $[0,1]$, ϕ^k is the piecewise linear real valued function on $[0,1]$ determined by $\phi^k(0) = 0$, $\phi^k(t_j) = x_j$, $j = 1,\ldots,k$, and Z_k is a normalization constant. The mesh of the partition is assumed to go to zero in the limit. If one takes the limit separately of the normalization constant, the integrand and the k dimensional Lebesgue measure then one obtains the heuristic equation

(2.2) $$\int_W f(\phi)\, dw(\phi) = Z^{-1} \int_W f(\phi) \exp\left[-\frac{1}{2}\int_0^1 \dot{\phi}(t)^2\, dt\right] \mathcal{D}\phi$$

where $\mathcal{D}\phi$ is to be thought of as infinite dimensional Lebesgue measure, customarily written $\mathcal{D}\phi = \Pi_{t\epsilon[0,1]}\, d(\phi(t))$. From the known behavior of the sample paths of a Wiener process we have $\int_0^1 \dot{\phi}(t)^2\, dt = +\infty$ for a.e. $\phi[w]$. Moreover since $Z_k = \Pi_{j=1}^k (2\pi(t_j - t_{j-1}))^{1/2}$, we have $Z = 0$. Thirdly $\mathcal{D}\phi$ by itself has no simple, clear cut meaning. Nevertheless the expression (2.2) has much heuristic value for computations, suggests correct theorems, and is the customary way of describing the Wiener integral in much of the physics literature.

The generalziation of (2.2) which corresponds to the quantized free scalar field in n space dimensions is similarly given by the informal expression

(2.3) $$dm(\phi) = Z^{-1} \exp\left[-\frac{1}{2}\int_{R^d} (|\text{grad}\, \phi(x)|^2 + \phi(x)^2)dx\right] \mathcal{D}\phi$$

where $d = n+1$ and $\mathcal{D}\phi$ is an infinite dimensional Lebesgue measure. This expression may be given a meaning as a probability measure on $\mathcal{S}'(R^d)$ by using a lattice construction of m analgous to that in equation (2.1) [9]. But it is also instructive to interpret (2.3) with the help of Minlos' Theorem. By an informal integration by parts in the exponent in (2.3) we may write

$$(2.4)\qquad dm(\phi) = Z^{-1} \exp\left[-\tfrac{1}{2}((1-\Delta)\phi,\phi)_{L^2(R^n)}\right] \mathcal{D}\phi$$

By analogy with the finite dimensional Gaussian Fourier transform formula

$$Z^{-1}\int_{R^k} e^{i(x,y)} \exp\left[-\tfrac{1}{2}(Ax,x)\right]dx = \exp\left[-\tfrac{1}{2}(A^{-1}y,y)\right],$$

which is valid if A is a strictly positive definite matrix on R^k and $Z := (2\pi)^{k/2}(\det A)^{-1/2}$ is the normalization constant, we therefore define m as the unique probability measure on $\mathcal{S}'(R^d)$ whose Fourier transform is given by

$$(2.5)\qquad \int_{\mathcal{S}'(R^d)} e^{i\phi(f)}\, dm(\phi) = \exp\left[-\tfrac{1}{2}((1-\Delta)^{-1} f,f)_{L^2(R^d)}\right]$$

Here f is in $\mathcal{S}(R^d)$. The existence and uniqueness of m is assured by Minlos' theorem. The generalized random field given by the map $f \to \phi(f)$ from $\mathcal{S}(R^d)$ to measurable functions on $(\mathcal{S}'(R^d), m)$ has some remarkable Markovian properties and is invariant under the Euclidean group of R^d [12]. Just as in the case of Wiener measure the exponent in (2.3) is infinite for m-almost all sample functions ϕ. But unlike the Wiener process the map $\phi \to \phi(x)$ is, for fixed x, not a well defined measurable function [5] with respect to m. That is to say m-almost every ϕ is not a measurable function on R^d if $d \geq 2$.

It is easy to describe now the Yang-Mills measure in the informal spirit of equations (2.2) and (2.3). Let G be a closed subgroup of the group $U(N)$ of $N \times N$ unitary matrices. Denote by $\mathcal{G}$ its Lie algebra, which we identify with a space of anti-Hermitian $N \times N$ matrices. The Yang-Mills measure, μ, for G in d space-time dimensions is defined (in so far as it can be defined) by the equation

$$(2.6)\qquad d\mu(A) = Z^{-1} \exp\left[-\frac{1}{2c^2}\|F\|^2\right] \mathcal{D}A$$

Here A is a (generalized) 1-form with values in $\mathcal{G}$; that is, $A = \sum_{i=1}^{d} A_i(x)dx^i$ with $A_i \in \mathcal{S}'(R^d;\mathcal{G})$. F is its curvature: $F = dA + A \wedge A$. In coordinates,

$$(2.7)\qquad F_{ij}(x) = \partial A_j(x)/\partial x^i - \partial A_i(x)/\partial x^j + [A_i(x), A_j(x)].$$

As usual, [,] denotes the commutator. Further,

$$(2.8)\qquad \|F\|^2 = -\int_{R^d} \text{trace}\ (\sum_{i<j} F_{ij}(x)^2)\ dx.$$

Finally $\mathcal{D}A = \Pi_{j=1}^d \ \Pi_{x\varepsilon R^d}\ d(A_j(x))$ is an infinite dimensional "Lebesgue measure" and Z is a normalization constant. c is a strictly positive real number.

The expression (2.6) differs from (2.3) in at least two significant respects. First, the exponent, $\|F\|^2$, is not quadratic in A but quartic, if G is not commutative, since F itself is quadratic in A by (2.7). Second, $\|F\|^2$ is a degenerate form as a function of A whether G is commutative or not. This degeneracy can be understood most easily if G is commutative. Consider the important case $G = U(1)$ (the circle group). This choice of G corresponds to electromagnetism. In this case $F = dA$, and we see that if A is exact, i.e., $A = d\lambda$ for some purely imaginary generalized function λ on R^d, then $F = dA = d^2\lambda = 0$. Thus the infinite dimensional space of exact 1-forms A is in the kernel of the quadratic form $\|dA\|^2$. Without being precise (for the moment) about what infinite dimensional space A runs over, we may understand the degeneracy problem by analogy with the measure $Z^{-1}\exp[-u^2/2]du\ dv$ in the plane. The exponent is a degenerate quadratic form in the two real variables u,v. Consequently the measure is a product of a one dimensional Gauss measure times a one dimensional Lebesgue measure. The expression (2.6) for μ may similarly be regarded as a product of an infinite dimensional Gauss measure (analogous to (2.3)) times an infinite dimensional (meaningless) Lebesgue measure. The customary interpretation of μ in this (commutative) case amounts, in effect, to ignoring the Lebesgue measure factor and interpreting μ as the Gaussian measure factor on the quotient space obtained by dividing out the space, $\mathcal{C}$, over which A runs by the kernel of the quadratic form. In our two dimensional example this means giving a meaning to

$$(2.9)\qquad Z^{-1}\int \psi(u,v)\ e^{-u^2/2}\ du\ dv$$

for functions ψ which depend only on u. We could simply ignore the dv integral! What amounts to the same thing but captures the spirit of the customary procedure in the physics literature is this: observe that

$$(2.10) \qquad \iint_{R^2} \psi(u)\ (m/(2\pi))\ \exp[-(u^2 + m^2v^2)/2]\ du\ dv$$

is independent of m for $m > 0$ and is the integral of ψ with respect to a probability measure for each m. At a very informal level we may regard (2.9) with $Z = \infty$ as the limit of (2.10) as $m \downarrow 0$. We view (2.10) as defining a probability measure on the quotient space, $R^2/(v\text{ axis})$. The corresponding procedure in the infinite dimensional case of interest consists in adding to $\|F\|^2$ a so called "gauge fixing term" $m^2\|\operatorname{div} A\|^2_{L^2(R^d)}$ which renders the exponent in (2.6) nondegenerate. As in the case of (2.3) Minlos' theorem can now be applied to make sense out of the thus modified version of (2.6) as a probability measure on $\mathcal{G}:=R^d \otimes \mathcal{S}'(R^d)$. In particular if ρ is a 2-form on R^d with coefficients in $\operatorname{Re} \mathcal{S}(R^d)$ and we put

$$F(\rho) = \sum_{i,j} \int_{R^d} (dA)_{ij}(x)\ \rho_{ij}(x)dx$$

(the right side has only symbolic meaning since each A_j is in $\mathcal{S}'(R^d)$) then the functional $A \to F(\rho)$ is clearly invariant under translation by elements of the kernel of $\|dA\|^2_{L^2(R^d)}$. An informal Gaussian computation analogous to that leading to (2.5) gives

$$(2.11) \qquad E_\mu(e^{F(\rho)}) = \exp[-(c^2/2)((-\Delta)^{-1}\ d^*\rho,\ d^*\rho)_{L^2(R^d)^d}]$$

The inner product on the right is on 1-forms and d^* is the usual adjoint of d ($d^* = \operatorname{div}$). Note that the right side of (2.11) is independent of the parameter m. We may now convert the preceding heuristics into a precise definition.

<u>Definition</u>. The <u>free Euclidean electromagnetic field</u> is the purely imaginary random process $\rho \to F(\rho)$, indexed by the set $\Lambda^2(R^d) \otimes \mathcal{S}(R^d)$ of 2-forms ρ with coefficients in $\operatorname{Re} \mathcal{S}(R^d)$ and whose characteristic functional is given by the right side of (2.11).

Minlos' theorem ensures the existence of a unique (Gaussian) probability measure μ on $\Lambda^2(R^d) \otimes \mathcal{S}'(R^d)$ satisfying (2.11). It is clear from (2.11) that $F(\rho) = 0$ a.e., if $d^*\rho = 0$. Thus μ may be regarded as a measure on $\mathcal{M} \equiv \{F \in \Lambda^2(R^d) \otimes \mathcal{S}'(R^d) : F(\rho) = 0$ whenever $d^*\rho = 0\}$. It is conceptually important for making the transition to the non-abelian case to observe that this set is the closure of the image, under the map induced by $A \to dA$, of the quotient space $\mathcal{G}^\infty/\{A \in \mathcal{G}^\infty : dA = 0\}$ where

$$\mathcal{G}^\infty = \{A \in R^d \otimes \mathcal{S}'(R^d) : A_j \in C^\infty(R^d),\ j = 1,\dots,d\}$$

The lesson to be drawn from the preceding discussion of the kernel problem in the commutative case can be described in the non abelian case (general Yang-Mills field) as follows. Let g be in $C^\infty(R^d;G)$ where G is now a closed connected subgroup of $U(N)$. Define

$$A^g(x) = g(x)^{-1} A(x)\, g(x) + g(x)^{-1}\, dg(x) \tag{2.12}$$

As is well known [10] the map $A \to A^g$ is the transformation law for a connection form, A, on R^d under the global change of global chart determined by the transition function $g(\cdot)$ in the vector bundle $R^d \times \mathbb{C}^N \to R^d$. Moreover the induced action on the curvature F is $F \to F^g$ where $(F^g)_{ij}(x) = g(x)^{-1} F_{ij}(x)\, g(x)$. Consequently $\|F^g\|^2 = \|F\|^2$. The gauge transformation $A \to A^g$ is the non abelian generalization of the translation $A \to A + d\lambda$ which, as we saw, leaves the quadratic form $\|F\|^2$ invariant in case $G = U(1)$. (Take $g(x) = e^{\lambda(x)}$ where $\lambda : R^d \to iR \equiv$ Lie algebra of $U(1)$). Moreover the Lebesgue measure $\mathcal{D}A$ in (2.6) which is (informally) an infinite product of Lebesgue measure on $\mathcal{G}$ is (informally) invariant under the map $A \to A^g$ because for each x in R^d $(A^g)_j(x)$ is a rotation of $A_j(x)$ in plus a translation. Because of the invariance of $\|F\|^2$ under these rotations plus translations of the infinite dimensional space of A's the kernel problem described above in the abelian case persists in the non abelian case. It is expected therefore that if the expression (2.6) can be given a meaning as a measure (and this has not been done yet if $d \geq 3$) then the measure will live on some quotient space (or its completion) analogous to that for the abelain case, $U(1)$. To be somewhat more precise (and speculative, by necessity) let

$$\mathcal{G}^\infty = \{\text{1-forms } A = \sum_{j=1}^{d} A_j(x)\,dx^j : A_j \in C^\infty(R^d;\mathcal{G})\}$$

If g is in $C^\infty(R^d;G)$ then g acts on $\mathcal{C}^\infty$ by (2.12). Let $\mathcal{M}^\infty = \mathcal{C}^\infty/C^\infty(R^d;G)$. Then $\mathcal{M}^\infty$ is a smooth non abelian analog of the quotient space discussed at the end of the last paragraph. Because of the smoothness imposed on elements of $\mathcal{C}^\infty$ the quotient space is too small to support the hoped for measure, μ. Rather, one expects that if μ exists as a measure at all then it will live on some kind of distribution completion of $\mathcal{M}^\infty$ analogous to $\mathcal{M}$. But even the problem of finding reasonable functions on $\mathcal{M}^\infty$ which could be effectively used to define the completion and the relevant σ-fields is in a primitive state of understanding at the present time. See [8] for further discussion of this point. In the meanwhile the geometry and topology of $\mathcal{M}^\infty$ has been investigated [1,2,11,14,15] when R^d is replaced by a compact manifold. In spite of the lack of a clear cut formulation of just what (2.6) ought to mean or where it should live (but see [8]) we shall look now at a construction scheme analogous to Wiener's construction method (2.1).

§3. The Wilson lattice approximation. K. Wilson [16] described a measure which, in a compelling though necessarily heuristic sense, appears to be an approximation to the expression (2.6) and may provide a mechanism for proving the existence of a measure interpretable as (2.6) for any closed subgroup G of $U(N)$. Let $\varepsilon > 0$. Replace R^d by the lattice $L := \varepsilon\mathbb{Z}^d$ of lattice spacing ε. For x in $\varepsilon\mathbb{Z}^d$ replace $A_j(x)$ by an element $g_j(x)$ in G. We think of $g_j(x)$ as being approximately the operator of parallel translation along the straight line segment from x to $x+\varepsilon e_j$ determined by the connection form A. That is, $g_j(x) \cong \exp(\varepsilon A_j(x))$. Here $e_1,\dots,e_d$ is the standard O.N. basis of R^d. Denote by dg normalized Haar measure on the compact group G. Replace the infinite dimensional Lebesgue measure $\mathcal{D}A$ which appears in (2.6) by

$$(3.1) \qquad d\nu(g) = \Pi_{j=1}^{d} \; \Pi_{x\in L} \; d(g_j(x))$$

The right side is an infinite product of probability measures and is therefore a well defined probability measure on $G^L \times\dots\times G^L$ (d factors - one for each coordinate direction). On the left side of (3.1) g denotes a function on $\{1,\dots,d\}\times L$ whose values appear on the right.

Next we shall describe the lattice version of the density which appears in (2.6) and then we shall discuss the heuristic reason for believing that as $\varepsilon \downarrow 0$ the resulting well defined measure "converges" to (2.6). Pick a point x in L and a pair of coordinate directions $j < k$. Given a configuration $g = \{g_i(x)\}$ as in (3.1) associate a kind of approximate exponentiated curvature as follows. Let

$$g_{jk}(x) = g_k(x)^{-1} g_j(x + \varepsilon e_k)^{-1} g_k(x + \varepsilon e_j) g_j(x). \tag{3.2}$$

$g_{jk}(x)$ clearly represents parallel translation around a little square in the jk plane with lower left hand corner at x. The usual geometric definition of curvature suggests that for small ε $g_{jk}(x)$ is approximately $\exp[\varepsilon^2 F_{jk}(x)]$ (which is correct if A is smooth and F is its curvature).

$$g_{jk}(x) \tilde{=} \exp [\varepsilon^2 F_{jk}(x)]. \tag{3.3}$$

A reader willing to grant the validity of (3.3) in some as yet unspecified sense of approximation as $\varepsilon \downarrow 0$ (and bearing in mind that the typical $A_j(x)$ producing $g_j(x)$ will be far from smooth) will certainly grant also that up to order ε^4 we may write

$$\begin{aligned} \text{Re trace } g_{jk}(x) &= \text{Re trace } (I+\varepsilon^2 F_{jk}(x) + \frac{\varepsilon^4}{2} F_{jk}(x)^2 + \ldots) \\ &= N + \frac{\varepsilon^4}{2} \text{ trace } (F_{jk}(x)^2) + O(\varepsilon^6) \end{aligned} \tag{3.4}$$

We have used Re trace $F_{jk}(x) = 0$. (Remember that $F_{jk}(x)$ is in $\mathcal{G}$, which is contained in the set of anti Hermitian $N \times N$ matrices). Consequently

$$\begin{aligned} &\sum_{j<k} \sum_{x\in L} (\text{Re trace } g_{jk}(x) - N)\, \varepsilon^{d-4} \\ &= (1/2) \sum_{j<k} \sum_{x\in L} \varepsilon^d \text{ trace } (F_{jk}(x)^2) + \varepsilon^2 \sum_{j,k} \varepsilon^d R(x,\varepsilon) \end{aligned} \tag{3.5}$$

where $R(x,\varepsilon)$ is a remainder term. Under the (inapplicable) assumption that A, and hence F, have coefficients in $C_c^\infty(R^d)$ the first term on the right converges to

$$(1/2) \sum_{j<k} \int_{R^d} \text{trace } (F_{jk}(x)^2)\, dx$$

while the second term converges to zero. This argument shows that the left side of (3.5) is an approximation to the exponent in the density which appears in (2.6), at least if A is smooth and compactly supported (a circumstance which can be expected to hold with μ probability zero).

Let μ_ε denote the probability measure on $G^L\times\ldots\times G^L$ (d factors) given by

$$(3.6) \qquad d\mu_\varepsilon(g) = Z^{-1} \exp\Big[\frac{\varepsilon^{d-4}}{c^2} \sum_{|x|<\varepsilon^{-1}} \text{Re trace } g_{jk}(x)\Big]\, d\nu(g)$$

where Z is the normalization constant. μ_ε is clearly an honest probability meausre. We have seen that (at a highly informal level) the exponential factor in (3.6) converges to the exponential factor in (2.6). (The omitted constant term N can be absorbed into Z.) Moreover Lebesgue measure on $\mathcal{G}$ and Haar measure on G are approximately equal on sets which are close to the origin in $\mathcal{G}$, so that if $g = e^{\varepsilon A}$ then $dg \doteq \text{constant}\cdot dA$ for A in $\mathcal{G}$ and ε small, where dA = Lebesgue measure on $\mathcal{G}$. For example if $G = U(1)$ then Haar measure expressed in terms of $e^{i\theta}$ is $(2\pi)^{-1}\, d\theta$. We may therefore view the measure ν as a normalized approximation to the factor $\mathcal{D}A$ of (2.6). Combining this with the preceding discussion of the exponential factor suggests that μ_ε is in some sense an approximation to μ. This is the lattice approximation of K. Wilson [16].

Closer inspection of the preceding argument for convergence of μ_ε as $\varepsilon \downarrow 0$ shows that it hinges on the condition that for each j,k and x the distribution of $g_{jk}(x)$ in G induced by μ_ε converges to the point mass at the identity of G. This captures somewhat more accurately the true meaning of the equation (3.3). If $d = 2$ or 3 this necessary condition holds. If $d=2$ then the measure μ in (2.6) can be given a meaning and the convergence of μ_ε to μ undoubtedly holds, but complete details are not in print (see [3,4].) If $d=3$ a proof of convergence of μ_ε has been published only in case $G = U(1)$ [7]. Convergence is in the sense of convergence of characteristic functionals as well as in the sense of convergence of moments. The theorem stated below makes this more explicit. If $d \geq 4$ then the distribution of $g_{jk}(x)$ no longer converges to the point mass at the identity of G as $\varepsilon \downarrow 0$. (The factor ε^{d-4}/c^2 in the exponent in (3.6) is the key ingredient in determining this behavior). Therefore if $d \geq 4$ there isn't a good heuristic basis for believing that μ_ε converges to something that deserves to be represented by (2.6). But the case of real physical interest is $d=4$. It is widely believed that in this case and with $G = SU(N)$ one should let the coupling constant c in (3.6) go to zero along with ε at just the proper rate that the μ_ε will then converge to a (perhaps generalized) measure which represents the true meaning of (2.6), with $c = 0$. The circle of ideas involved with letting $c(\varepsilon) \downarrow 0$ goes under the name assymptotic freedom.

The following theorem shows that in case $d=3$ and $G = U(1)$ the preceding heuristic discussion is justified. Write

$$f_{jk}^{(\varepsilon)}(x) = \varepsilon^{-2} \operatorname{Im} g_{jk}(x) \quad .$$

Then in view of (3.3) $f_{jk}^{(\varepsilon)}(x)$ should be regarded as the lattice approximation to $F_{jk}(x)$.

<u>Theorem 1</u>. Let $d=3$ and $G = U(1)$. Let ρ be a real valued 2-form on R^3 of class C_c^∞. Put

$$F_\varepsilon(\rho) = \sum_{j<k} \sum_{x\in L} \varepsilon^3 \rho_{jk}(x) f_{jk}^{(\varepsilon)}(x)$$

Then $\int \exp[iF_\varepsilon(\rho)]d\mu_\varepsilon(g)$ converges as $\varepsilon \downarrow 0$ to the right hand side of (2.11) if ρ is exact.

<u>Remark</u>. The requirement that ρ be exact in the statement of the theorem is related to the fact that the lattice approximation introduces spurious magnetic monopoles. That is, the equation $dF = 0$, which holds for the free Euclidean electromagnetic field by (2.11), fails on the lattice and must be shown to hold in the limit as $\varepsilon \downarrow 0$. This is explained in more detail in [7] where only partial results on this point are obtained.

REFERENCES

[1] Asorey, M., and Mitter, P.K., Regularized, continuum Yang-Mills process and Feynman-Kac functional integral, Commun. Math. Phys. 80,(1981), 43-58.

[2] Asorey, M., and Mitter, P.K., On geometry, topology and θ-sectors in a regularized quantum Yang-Mills theory, Preprint Cern/TH 3424 (1982).

[3] Balian, R., Drouffe, J.M., and Itzykson, C., Phys. Rev. D10 (1974) 3376; D11 (1975), 2098 and 2104.

[4] Bralić, N.E., Exact computation of loop averages in two-dimensional Yang-Mills theory, Phys. Rev. D 22 (1980) 3090-3103.

[5] Colella, P., and Lanford, O., Sample field behavior for the free Markov random field, in Lecture Notes in Physics, Vol. 25 "Constructive Quantum Field Theory" Ed. G. Velo and A. Wightman, Springer 1973, p. 44-70.

[6] Glimm, J., and Jaffe, A., Quantum Physics, Springer-Verlag, New York, 1981.

[7] Gross, L., Convergence of $U(1)_3$ lattice gauge theory to its continuum limit, Commun. Math. Phys. 92 (1983), 137-162.

[8] Gross, L., A Poincaré Lemma for connection forms, J. Funct. Anal. 1985, to appear.

[9] Guerra, F., Rosen, L., and Simon, B., The $P(\phi)_2$ Euclidean quantum field theory as classical statistical mechanics, Ann. of Math. 101 (1975), 111-259.

[10] Kobayashi, S., and Nomizu, K., Foundations of differential geometry, Vol. 1,.Interscience Pub. Co. NY, 1963.

[11] Mitter, P.K., and Viallet, C.M., On the bundle of connections and the gauge orbit manifold in Yang-Mills theory, Commun. Math. Phys. 79 (1981), 457-472.

[12] Nelson, E., The free Markov field, J. Funct. Anal. 12 (1973), 211-227.

[13] Simon, B., The $P(\phi)_2$ Euclidean (quantum) field theory, Princeton Univ. Press, 1974.

[14] Singer, I.M., Some remarks on the Gribov ambiguity, Commun. Math. Phys. 60 (1978), 7-12.

[15] Singer, I.M., The geometry of the orbit space for non-abelian gauge theories Physica Scripta, 24 (1981) 817-820.

[16] Wilson, K.G., Confinement of quarks, Phys. Rev. D10 (1974), 2445-2459.

The Generalized Malliavin Calculus based on Brownian sheet and Bismut's Expansion for Large Deviation

Shigeo Kusuoka

Department of Mathematics
Faculty of Science
University of Tokyo

Introduction. The asymptotic behaviour of the transition probability density $p(t,x,y)$ of a diffusion process as $t \downarrow 0$ has been studied by a lot of analysts and probabilists. Especially probabilists have studied it from a view point of functional integral (e.g. Molchanov[7], Azencott[1],...). But they thought of the case only where the generator of the diffusion process is uniformly elliptic (though Gaveu[3] is exceptional).

Recently Bismut[2] gave a new viewpoint and showed an expansion formula. His idea is really natural, but it seems to us that his justification of his idea is too complicated. Actually he gave a natural conjecture in his paper and we thought that we could show his expansion formula in much larger cases than the case where he conjectured, only if we give a natural justification to his idea. The purpose of the present paper is to explain the strategy how one can refine and improve Bismut's results. Our main tool is generalized Malliavin calculus based on Brownian sheet whose basic idea of this calculus was first given in Malliavin [6]. This is a joint work with D. Stroock and the detail will be published in a series of papers.

1. Bismut's idea. Let us remind Bismut's idea in this section. The argument here is not rigorous, but it will make points clear. We will show our strategy in the last of this section. Now let us think of the stochastic differential equation:

$$\begin{cases} dX_t^\varepsilon = \varepsilon \sum_{i=1}^{d} V_i(X_t^\varepsilon)\, d\theta_t^i \\ X_0^\varepsilon = x_0 , \end{cases}$$

where $\varepsilon > 0$, V_i, $i = 1,\dots,d$, are C^∞ vector fields over R^N with $\sup_{x\ R^N} \left| \frac{\partial^\alpha}{\partial x^\alpha} V_i(x) \right| < \infty$ for any multi-index α and $\{ \theta_t^i;\ t \geq 0,\ i = 1,\dots,d \}$ is a d-dimensional Brownian motion. We assume

Hypothesis 0. $\mathrm{Lie}[V_1,\dots,V_d](x_0) = R^N$.

Then the probability distribution of X_1^ε has a smooth density $p(y) = p(s,x_0,y)$, $s = \varepsilon^2$. Since $P[\ X_s^1 \in dy\ ] = P[\ X_1^\varepsilon \in dy]$ $= p(s,x_0,y)dy$, $p(s,x_0,y)$ is the transition probability density function of a diffision process whose generator is $L = \frac{1}{2} \sum_{i=1}^{d} V_i^2$.

Our concern is in the asymptotic behavior of $P(s,x_0,y)$, $s \downarrow 0$. In other words, we want to know the asymptotic behavior of $P[\ X_1^\varepsilon \in dy\ ]$ as $\varepsilon \downarrow 0$.

Let $\Theta = \{\ \theta : [0,1] \to R^N;\ \text{continuous}\ ,\ \theta(0) = 0\ \}$. Then there is a measurable function $F : \Theta \to R^N$ with $X_1^\varepsilon(\theta) = F(\theta)$. We may think that $X_1^\varepsilon(\theta) = F(\varepsilon\theta)$ with probability one. Let $M = \{\ \theta \in \Theta;\ F(\theta) = y\ \}$. We assume also the following.

Hypothesis 1. There exists a unique h_0 such that $\|h_0\|_H = \inf\{\ \|\theta\|_H;\ \theta \in \Theta\ \} < \infty$. Here $\|\theta\|_H^2 = \int_0^1 \|\dot\theta(t)\|^2\, dt$. We denote by H a Hilbert space $\{\ \theta \in \Theta;\ \|\theta\|_H < \infty\ \}$ with a Hilbert norm $\|\ \ \|_H$.

Hypothesis 2. $F|_H : H \to R^N$ is smooth, and its Fréchet derivative

$DF(h_0): H \to R^N$ at h_0 has full rank N.

Let N_0 denote Image $DF(h_0)^* \subset H$ and T_0 denote ker $DF(h_0) \subset H$. Then it is easy to see that $h_0 \in N_0$. Now we start to think of the asymptotic behavior of $p(s,x_0,y)$ for fixed y.

$$
\begin{aligned}
& p(s,x_0,y) \\
= \; & P[\, F(\varepsilon\theta) \in dy \,]/dy \\
= \; & \left(\frac{1}{2\pi}\right)^{\dim H/2} \int_H \delta(\, F(\varepsilon\theta) - y \,) \exp\left(- \frac{1}{2} \|\theta\|_H^2 \right) d\theta \\
= \; & \left(\frac{1}{2\pi}\right)^{\dim H/2} \int_{T_0} du \int_{N_0} \frac{dz}{\varepsilon^N} \delta(\, F(h_0+\varepsilon u+z) - y \,) \\
& \qquad \exp\left(- \frac{1}{2}\left(\|u\|_H^2 + \frac{1}{\varepsilon^2} \|h_0+z\|_H^2 \right) \right).
\end{aligned}
$$

Here we set $\theta = u + \frac{1}{\varepsilon}(h_0+z)$ and $d\theta$, dz, du are flat measures on H, N_0, T_0 respectively. Since only neighborhood of h_0 contributes to the principal part of $p(s,x_0,y)$, $s\downarrow 0$, and $DF(h_0)$ has full rank we may assume that there is a smooth function $z : T_0 \to N_0$ such that $M \cap$ a neiborhood of h_0 in $H = \{\, h_0 + u + z(u);\ u \in$ a neiborhood of the origin in $T_0 \,\}$. Then

$$
\begin{aligned}
& p(s,x_0,y) \\
\sim \; & \left(\frac{1}{2\pi\varepsilon^2}\right)^{N/2} \exp\left(- \frac{1}{2\varepsilon^2} \|h_0\|_H^2 \right) \\
& \times \left(\frac{1}{2\pi}\right)^{\dim T_0/2} \int_{T_0} g(\varepsilon,u) \exp\left(- \frac{1}{\varepsilon^2} f(\varepsilon u) \right) \\
& \qquad \exp\left(- \frac{1}{2} \|u\|_H^2 \right) du ,
\end{aligned}
$$

where $g(\varepsilon,u) = \det[\,(DF(h_0+u+z(\varepsilon u))|_{N_0})(DF(h_0+u+z(\varepsilon u))|_H{}^*)\,]^{-1/2}$

and $f(u) = (z(u),h_0)_H + \frac{1}{2} \|z(u)\|_H^2$.

It is easy to see that $f(0) = 0$, $Df(0) = 0$ and

$D^2f(0)(k_1,k_2) = (\, DF(h_0)^*(DF(h_0)DF(h_0)^*)^{-1}DF(h_0)(k_1,k_2),\ h_0 \,)_H$.

Now let us assume the following.

<u>Hypothesis 3.</u> $D^2f(0) + I_{T_0}$ is strictly positive definite on T_0.

Then from Schilder's theorem, we see that

$$p(s,x_0,y) \sim \left(\frac{1}{2\pi s}\right)^N \exp\left(- \frac{1}{2s} \|h_0\|_H^2\right) \sum_{k=0}^{\infty} a_k s^k ,$$

where $a_0 = \det[DF(h_0)\, DF(h_0)^*]^{-1/2} \det[I_{T_0} + D^2f(0)]^{-1/2}$.

This is Bismut's expansion formula.

It is not easy to justify the above argument. The main difficulties are the following.

(1) How one can give the precise meaning for $F(\varepsilon u)$, $z(u)$ and so on in general? Note that the probability measures on Θ induced by $\varepsilon\theta$ and the Wiener measure, $\varepsilon > 0$, are mutually singular.

(2) How one can justify Schilder type theorem for such a non-continuous functional on Θ? Taylor's expansion does not apply in this case.

(3) Since we restrict the functional F on T_0 , we have to be involved in noncausal stochastic integral. But noncausal stochastic integral is not convinient tool. Can one avoid it?

Bismut[2] considered concrete functionals, solutions of S.D.E., and so he did not have to think of the definition of $F(\varepsilon\theta)$. He did not avoid to think of noncausal stochastic integral. He made effort in giving the definition for $z(u)$ and in showing Schilder type theorem. We take completely different strategy from his. First we think of d-dimensional Brownian sheet $W^s(t) = W(s,t) = \{W^i(s,t);\ i=1,\dots,d\}$ and think of functionals of W^s. For example, let us think of S.D.E.

$$\begin{cases} dX(t;s,W^s) = \sum_{i=1}^{d} V_i(X(t))\, dW^s{}_i(t) + s\, V_0(X(t))dt \\ X(0;s,W^s) = x \end{cases}$$

Then functionals $X(1;s,W^s)$, $s \geqq 0$, are our object. Since the probabilty $P[\ X(1;s,W^s) \in dy\ ]$ is equal to the transition probability $p(s,x,dy)$ of the diffusion process on R^N with a generator $L = \frac{1}{2}\sum_{i=1}^{d} V_i^{\,2} + V_0$, our concern is in the asymptotic behavior of functionals as $s \downarrow 0$. We will show Bismut's expansion formula for general class of functionals in the following way.

Step 1. We construct stochastic analysis including generalized Malliavin's calculus for functionals of a Brownian motion on an abstract Wiener space. Then we will get stochastic Taylor expansion.

Step 2. We show an abstract Schilder type theorem by relying on Wentzel-Freidlin type argument and stochastic Taylor expansion.

Step 3. We show that our class of functionals are robust even if we split the Hilbert space and the Brownian motion as in the case where $H = T_0 \oplus N_0$. Then we will get Bismut's expansion formula automatically.

Finally let us make some remarks on the advantage of our method. First of all, we can think of various kinds of asymptotic problems at the same time. For example, we may think of the asymptotic behavior of $X(1;s,W^s)$ where $X(t;s,W^s)$ satisfies

$$\begin{cases} dX(t;s,W^s) = \sum_{i=1}^{d} V_i(X(t))\, dW^s{}_i(t) + V_0(X(t))dt \\ X(0;s,W^s) = x \quad . \end{cases}$$

We can handle Ito (non-Markovian) process. We can also think of the case where the set $\{\ h \in M\ ;\ \|h\|_H = \inf\{\ \|\theta\|_H;\ \theta \in M\ \}\}$ makes a finite dimensional manifold, but we will not discuss this case in the present paper.

2. Generalized Malliavin Calculus. Let B be a separable real

Banach (or Fréchet) space and H be a separable Hilbert space for which H is continuously and densely embedded in B. We identify the dual space H^* of H with H itself. Then the dual space B^* of B can be regarded as a dense subspace of H. Let μ_s, $s \geqq 0$, be Gaussian measures on B such that

$$\int_B \exp(\sqrt{-1}\ {}_B\langle z,u\rangle_{B^*})\ \mu_s(dz) = \exp(- \frac{s}{2} \|u\|_H^2) \quad \text{for any } u \in B^*.$$

We will write W or W_B for the totality of continuous functions from $[0,\infty)$ into B. Let $P = P_{(B,H,\mu_s,s\geqq 0)}$ be a probability measure given by

$$P[\ w(t_1) \in A_1,\ w(t_2)-w(t_1) \in A_2,\dots,\ w(t_n)-w(t_{n-1}) \in A_n\ ]$$
$$= \prod_{i=1}^{n} \mu_{t_i - t_{i-1}}(A_i)$$

for any $n \geqq 1$, $0 = t_0 \leqq t_1 < t_2 < \dots < t_n < \infty$, $A_i \in \mathcal{B}(B)$, $i = 1,\dots,n$. We will call the triple (H,B,P) an abstract Wiener process. This object was first studied by Gross[4].

For any separable real Hilbert space E, we set

$$FC^\infty_\uparrow([0,\infty)\times B;E)$$
$= \{\ f \in C^\infty([0,\infty)\times B;E)$; there are $n \geqq 1$, a continuous linear functional $A : B \to R^n$ and $\tilde{f}$ $C^\infty_\uparrow(R^{1+n};E)$ such that $f(s,z) = \tilde{f}(s,Az)$, $(s,z) \in [0,\infty)\times B\ \}$. Here

$C^\infty_\uparrow(R^n;E) = \{\ f \in C^\infty(R^n;E)$; for any multi-index α, there are $C_\alpha < \infty$ and $\nu_\alpha < \infty$ with $\left|\frac{\partial^\alpha}{\partial x^\alpha} f(x)\right|_E \leqq C_\alpha (1+|x|)^{\nu_\alpha}$, $x \in R^n\ \}$.

We define three operations on $FC^\infty_\uparrow([0,\infty)\times B;E)$ by the following. For any $f \in FC^\infty_\uparrow([0,\infty)\times B;E)$, we define Df $FC^\infty_\uparrow([0,\infty)\times B;\mathcal{H}(E))$ by

$$Df(s,z)(h) = \lim_{\tau\to 0} \frac{1}{\tau}\ (\ f(s,z+\tau h) - f(s,z)\ ),\ (s,z) \in [0,\infty)\times B\ .$$

Here $\mathcal{H}(E)$ denotes the Hilbert space consisting of all

Hilbert-Schmidt operators from H into E.

For any $f \in FC^{\infty}_{\uparrow}([0,\infty)\times B;E)$, we define $\mathcal{L}f \in FC^{\infty}_{\uparrow}([0,\infty)\times B;E)$ by

$$\mathcal{L}f(s,z) = \frac{1}{2}\{ s\cdot \mathrm{trace}_H D^2 f(s,z) - \lim_{\tau\to 0} \frac{1}{\tau}(f(s,(1+\tau)z) - f(s,z)) \},$$

$(s,z) \in [0,\infty)\times B$. Here $\mathrm{trace}_H D^2 f(s,z) = \sum_{i=1}^{\infty} D^2 f(s,z)(h_i,h_i)$ and $\{h_i\}_{i=1}^{\infty}$ is an arbitrary orthonormal basis in H.

For any $f \in FC^{\infty}_{\uparrow}([0,\infty)\times B;E)$, we define $\mathcal{A}f \in FC^{\infty}_{\uparrow}([0,\infty)\times B;E)$ by

$$f(s,z) = \frac{\partial}{\partial s} f(s,z) + \frac{1}{2}\mathrm{trace}_H D^2 f(s,z), \quad (s,z) \in [0,\infty)\times B .$$

Then it is easy to see the following.

<u>Proposition 2.1.</u> For any $f \in FC^{\infty}_{\uparrow}([0,\infty)\times B;E)$,

$$f(s,h+w(s)) = f(0,h) + \int_0^s Df(\tau,h+w(\tau))\, dw(\tau) + \int_0^s \mathcal{A}f(\tau,h+w(\tau))\, d\tau$$

P-a.s.w for any $s \geqq 0$ and $h \in H$.

For each S, R > 0 and $1 < p < \infty$, we define a semi-norm $\| \ \|_{p,S,R;E}$ on $FC^{\infty}_{\uparrow}([0,\infty)\times B;E)$ by

$$\|f\|_{p,S,R;E} = \sup_{0\leqq s\leqq S} \ \sup_{h\ H,\ \|h\|_H\leqq R} E^P[\ \|f(s,h+w(s))\|_E^{\ p}]^{1/p} .$$

We define $\mathcal{G}^0(\mathcal{A};E)$ to be the completion of $FC^{\infty}_{\uparrow}([0,\infty)\times B;E)$ with respect to the semi-norms $\{ \| \ \|_{p,S,R;E} ;\ 1 < p < \infty,\ S,R > 0 \}$.

<u>Proposition 2.2.</u> For any $f \in FC^{\infty}_{\uparrow}([0,\infty)\times B;E)$ and $1 < p < \infty$, the map $(s,h) \to f(s,h+w(s))$ from $[0,\infty)\times H$ into $L^p(W\to E;dP)$ is completely continuous, i.e., this map is weakly continuous on any bounded subset of $[0,\infty)\times H$.

For each S, R > 0 and $1 < p < \infty$, we define a semi-norm $\| \ \|^{\mathcal{K}}_{p,S,R;E}$ on $FC^{\infty}_{\uparrow}([0,\infty)\times B;E)$ by

$$\|f\|^{\mathcal{K}}_{p,S,R;E} = \|f\|_{p,S,R;E} + \|Df\|_{p,S,R;\ (E)} + \|\mathcal{L}f\|_{p,S,R;E} + \|\mathcal{A}f\|_{p,S,R;E} .$$

We define $\mathcal{K}(\mathcal{A};E)$ to be the completion of $FC^{\infty}_{\uparrow}([0,\infty)\times B;E)$ with respect to the semi-norms $\{ \| \|^{\mathcal{K}}_{p,S,R;E} ; 1 < p < \infty, S,R > 0 \}$. Then we get the following two basic propositions.

Proposition 2.3. For any $f \in \mathcal{K}(\mathcal{A};E)$,

$$f(s,h+w(s)) = f(0,h) + \int_0^s Df(\tau,h+w(\tau))\, dw(\tau) + \int_0^s \mathcal{A} f(\tau,h+w(\tau))\, d\tau$$

P-a.s.w for any $(s,h) \in [0,\infty)\times H$.

Proposition 2.4. If $f,g \in \mathcal{K}(\mathcal{A};E)$ and if $f(s,w(s)) = g(s,w(s))$ for $ds\otimes P(dw)$-a.e.(s,w), then $f = g$ in $\mathcal{K}(\mathcal{A};E)$.

The following theorem is essentially due to Malliavin[6].

Theorem 2.5. For any f_1, f_2, $f_3 \in \mathcal{K}(\mathcal{A};R)$ and $\varphi \in C_0^{\infty}((0,\infty))$,

(1) $$E^P[\, f_3(s,w(s))\, (\, Df_1(s,w(s)),\, Df_2(s,w(s)\,)_H\,]$$
$$= -\, E^P[\, f_1(s,w(s))\, (\, Df_2(s,w(s)),\, Df_3(s,w(s)\,)_H\,]$$
$$-\, \frac{1}{s}\, E^P[\, f_1(s,w(s))\, f_3(s,w(s))\, \mathcal{L} f_2(s,w(s))\,]\, ,\quad s > 0\, ,\text{ and}$$

(3) $$\int_0^{\infty} ds\, \varphi(s)\, E^P[\, f_3(s,w(s))\, (\, \mathcal{A} f_1(s,w(s)) - \frac{1}{s}\mathcal{L} f_1(s,w(s))\,)\,]$$
$$= -\int_0^{\infty} ds\, \varphi(s)\, E^P[\, f_1(s,w(s))\, (\, \mathcal{A} f_3(s,w(s)) - \frac{1}{s}\mathcal{L} f_3(s,w(s))\,)\,]$$
$$-\int_0^{\infty} ds\, \varphi'(s)\, E^P[\, f_1(s,w(s))\, f_3(s,w(s))\,]\, .$$

Remark 2.6. D and $\mathcal{A} - \frac{1}{s}\mathcal{L}$ are differential operators of order one, and so satisfy the chain rule.

We define Frechet spaces $\mathcal{K}^n(\mathcal{A};E)$, $n \geqq 1$, inductively by the following.

$$\mathcal{K}^1(\mathcal{A};E) \underset{\text{def}}{=} \mathcal{K}(\mathcal{A};E).$$

$f \in \mathcal{K}^{n+1}(\mathcal{A};E)$ if $f \in \mathcal{K}(\mathcal{A};E)$ and if there are $g_1 \in \mathcal{K}^n(\mathcal{A};\mathcal{H}(E))$ and g_2, $g_3 \in \mathcal{K}^n(\mathcal{A};E)$ such that $g_1(s,w(s)) = Df(s,w(s))$, $g_2(s,w(s)) = \mathcal{L} f(s,w(s))$, and $g_3(s,w(s)) = \mathcal{A} f(s,w(s))$ for $ds\otimes P(dw)$-a.e.(s,w), and the semi-norms $\| \ \|^{(n+1)}_{p,S,R;E}$, $1 < p < \infty$, $S,R > 0$, on $K^{n+1}(\mathcal{A};E)$ are given by

$$\|f\|^{(n+1)}_{p,S,R;E} = \|f\|_{p,S,R;E} + \|g_1\|^{(n)}_{p,S,R;(E)} + \|g_2\|^{(n)}_{p,S,R;E} + \|g_3\|^{(n)}_{p,S,R;E} .$$

We also define a Fréchet space $\mathcal{G}(\mathcal{A};E)$ to be $\bigcap_{n=1}^{\infty} K^n(\mathcal{A};E)$.

The following is an improved version of the result of Malliavin[6].

Theorem 2.7. Let $F = (F^1,\dots,F^N) \in \mathcal{G}(\mathcal{A};R^N)$ and let

$$A_{ij}(s,w(s)) = (DF^i(s,w(s)), DF^j(s,w(s)))_H + (\mathcal{A} - \tfrac{1}{s}\mathcal{L})F^i(s,w(s))\cdot(\mathcal{A} - \tfrac{1}{s}\mathcal{L})F^j(s,w(s)),$$

$s > 0$, $i,j = 1,\dots,N$. If for any $1 < p < \infty$ and $0 < a < b < \infty$,

$\int_a^b ds\, E^P[\det(A_{ij}(s,w(s)))_{i,j=1,\dots,N}^{-p}] < \infty$, then for any $\varphi \in C_0^\infty((0,\infty))$, the signed measure ρ_φ on R^N given by $\int_{R^N} \psi(x)\, \rho_\varphi(dx) = \int_0^\infty ds\, \varphi(s)\, E^P[\psi(F(s,w(s)))]$, $\psi \in C_0(R^N)$, has smooth density.

3. Stochastic Taylor Expansion. For later use, we introduce some notation and give stochastic Taylor expansion formula. Let $\mathcal{B}_s$ be the sub-σ-algebra of $\mathcal{B}(W)$ generated by $w(\tau)$, $0 \leq \tau \leq s$. Given $\alpha \in \mathcal{O} = \{\phi\} \cup \bigcup_{\ell=1}^{\infty} \{0,1\}^\ell$, set

$$|\alpha| = \begin{cases} 0 & \text{if } \alpha = \phi \\ \ell & \text{if } \alpha \in \{0,1\}^\ell \end{cases},$$

$$[\alpha] = \{ i \quad \{1,\dots,\ell\} ; \alpha_i = 1 \}, \text{ and}$$

$$\|\alpha\| = \begin{cases} 0 & \text{if } \alpha = \phi \\ \ell + |\{ i \in \{1,\dots,\ell\} ; \alpha_i = 0 \}| & \text{if } \alpha \in \{0,1\}^\ell \end{cases}.$$

Also, if $\alpha = (\alpha_1,\dots,\alpha_\ell) \in \mathcal{O} \setminus \{\phi\}$, define

$$\alpha_* = \alpha_\ell \ , \quad \text{and}$$

$$\alpha' = \begin{cases} \phi & \text{if } |\alpha| = 1 \\ (\alpha_1,\dots,\alpha_{\ell-1}) & \text{if } |\alpha| \geqq 2 \end{cases} .$$

We write $\mathcal{H}^\ell(E)$ for $\underbrace{\mathcal{H}(\mathcal{H}\dots(\mathcal{H}}_{\ell}(E))\dots)$.

For any $f \in \mathcal{G}(\mathcal{A};E)$ and $\alpha \in \mathfrak{A}$, we define $\mathcal{D}^\alpha f \in \mathcal{G}(\mathcal{A};\mathcal{H}^{|[\alpha]|}(E))$ inductively by

$$\begin{cases} \mathcal{D}^\phi f = f \\ \mathcal{D}^{(0)} f = f, \ \mathcal{D}^{(1)} f = Df, \text{ and} \\ \mathcal{D}^\alpha f = \mathcal{D}^{(\alpha_*)}(\mathcal{D}^{\alpha'} f) . \end{cases}$$

For any $\alpha \in \mathfrak{A}$ and $\mathcal{B}_s$-adapted $\mathcal{H}^{|[\alpha]|}(E)$-valued process g_α with $\int_0^S ds\ E^P[\ \|g_\alpha(s,w)\|^2_{\mathcal{H}^{|[\alpha]|}(E)}\ ] < \infty$ for any $S > 0$, we define

$I^\alpha(g_\alpha) : [0,\infty)\times W \to E$ inductively by

$$\begin{cases} I^\phi(g_\phi) = g_\phi \ , \\ I^{(0)}(g_{(0)})(s,w) = \int_0^s g_{(0)}(\tau,w)\ d\tau \ , \\ I^{(1)}(g_{(1)})(s,w) = \int_0^s g_{(1)}(\tau,w)\ dw(\tau) \ , \text{ and} \\ I^\alpha(g_\alpha) = I^{(\alpha_*)}(\ I^{\alpha'}(g_\alpha)\ ) . \end{cases}$$

Then applying Ito's formula several times, we have the following.

<u>Proposition 3.1.</u> For any $f \in \mathcal{G}(\mathcal{A};E)$ and $n \geqq 0$,

$$f(s,h+w(s)) = \sum_{\|\alpha\| \leqq n} I^\alpha(\mathcal{D}^\alpha f(0,h))(s,w) + R_n(s,w) \ , \quad \text{where}$$

$$R_n(s,w) = \sum\{\ I^\alpha(g_\alpha) \ ;\ \|\alpha\|=n+1 \text{ or } \|\alpha\|=n+2, \alpha_*=0\ \} \text{ and}$$

$g_\alpha(s,w) = \mathcal{D}^\alpha f(s,h+w(s))$. Moreover there exists $C_{p,n}$ for any $1 < p < \infty$ such that $E^P[\ |R_n(s,w)|^p]^{1/p} \leqq C_{p,n}\ s^{\frac{1}{2}(n+1)}$, $0 \leqq s \leqq 1$.

<u>Remark 3.2.</u> For any $s \geqq 0$, the probability law of $\{\ I^\alpha(K_\alpha)(s,w);\ \alpha \in \mathfrak{A},\ K_\alpha \in \mathcal{H}^{|[\alpha]|}(E)\ \}$ coincides with the probability law of

$\{ s^{|\alpha|/2} I^{\alpha}(K_{\alpha})(1,w);\ \alpha \in \mathfrak{A},\ K_{\alpha} \in \mathcal{H}^{|[\alpha]|}(E) \}$.

Since our expression of terms which will appear in Bismut's expansion formula is different from Bismut's one, we will give a certain remark in the rest of this section to make the relation clear. Let f $FC_{\uparrow}^{\infty}([0,\infty)\times B;E)$ and $\varepsilon > 0$. Let $w^{\varepsilon}(s;w) = \varepsilon\, w(\varepsilon^{-2}s)$, w W and $s \geqq 0$. Then the probability law of $w^{\varepsilon}(\cdot,w)$ and $w(\cdot)$ are the same under P(dw). Therefore setting $s = \varepsilon^2$, we obtain

$$\begin{aligned} & f(\varepsilon^2;\varepsilon w(1)+h) \\ = \ & f(s,w^{\varepsilon}(s)+h) \\ = \ & \sum_{\|\alpha\| \leqq n} I^{\alpha}(\mathcal{D}^{\alpha} f(0,h))(s,w^{\varepsilon}) + R_n(s,w^{\varepsilon}) \\ = \ & \sum_{k=0}^{n} \varepsilon^k \sum_{\|\alpha\|=k} I^{\alpha}(\mathcal{D}^{\alpha} f(0,h))(1,w) + R_n(s,w^{\varepsilon}) \quad \text{for any } h \in H. \end{aligned}$$

Thus it is easy to see that

$$\frac{\partial^n}{\partial \varepsilon^n} f(\varepsilon^2,\varepsilon w(1)+h)\Big|_{\varepsilon=0} = \sum_{\|\alpha\|=n} I^{\alpha}(\mathcal{D}^{\alpha} f(0,h))(1,w) \quad \text{P-a.e.w}$$

for any integer n and $h \in H$. Therefore we see that

$$D^m\Big\{ \frac{\partial^n}{\partial \varepsilon^n} f(\varepsilon^2,\varepsilon w(1)+h)\Big|_{\varepsilon=0}\Big\} = \sum_{\|\alpha\|=n} I^{\alpha}(\mathcal{D}^{\alpha} D^m f(0,h))(1,w)$$

P-a.s. for any integer m, n, and $h \in H$. Hence we have the following.

<u>Proposition 3.3.</u> For any $f \in \mathcal{G}(\mathcal{L};E)$, $h \in H$ and $\varepsilon > 0$, $f(\varepsilon^2,\varepsilon\cdot+h) \in \mathcal{G}(\mathcal{L};E)$ and there exist $f_n \in \mathcal{G}(\mathcal{L};E)$, $n \geqq 0$, such that

(1) $f_n(w(1)) = \sum_{|\alpha|=n} I^{\alpha}(\mathcal{D}^{\alpha} f(0,h))(1,w)$ P-a.e.w, and

(2) $\| f(\varepsilon^2,\varepsilon\cdot+h) - \sum_{k=0}^{n} \varepsilon^k f_k \|_{p;E}^{(m)} = O(\varepsilon^{n+1})$ as $\varepsilon \downarrow 0$ for any $n,m \geqq 0$ and $1 < p < \infty$. Here we use the notation of Kusuoka-Stroock[5].

4. Abstract Schilder's expansion Theorem. In this section, we give an abstract vertion of Schilder's expansion theorem.

Definition 4.1. We say that an element f of $\mathcal{G}(\mathcal{L};E)$ is regular if there exists $\{f_n\}_{n=1}^{\infty} \subset FC_{\uparrow}^{\infty}([0,\infty)\times B;E)$ such that

$$\overline{\lim_{n\to\infty}}\ \overline{\lim_{s\downarrow 0}}\ s \log P[\ \| f(s,w(s)) - f_n(s,w(s)) \|_E > \varepsilon\ ] = -\infty$$

for any $\varepsilon > 0$. We say that an element f of $\mathcal{G}(\mathcal{L};E)$ is completely regular if $\mathcal{D}^{\alpha} f \in \mathcal{G}(\mathcal{L};\mathcal{H}^{|[\alpha]|}(E))$ is regular for every $\alpha \in \mathfrak{A}$.

Then we have the following.

Theorem 4.2. Let $f \in \mathcal{G}(\mathcal{L};E)$ be regular. Then

$$\underline{\lim_{s\downarrow 0}}\ s \log P[\ f(s,w(s)) \in G\ ] \geqq - \inf\{\ \frac{1}{2}\|h\|_H^2\ ;\ f(0,h) \in G\ \}$$

for any open subset G of E, and

$$\overline{\lim_{s\downarrow 0}}\ s \log P[\ f(s,w(s)) \in K\ ] \leqq - \inf\{\ \frac{1}{2}\|h\|_H^2\ ;\ f(0,h) \in K\ \}$$

for any closed subset K of E.

Theorem 4.3. Let $f \in \mathcal{G}(\mathcal{L};R)$ be completely regular and assume the following.

(1) There exists $\lambda > 1$ such that

$$\overline{\lim_{s\downarrow 0}}\ s \log E[\ \exp(\ \frac{\lambda}{s} f(s,w(s))\ ]\ < \infty\ .$$

(2) There exists a unique h_0 H satisfying

$$f(0,h_0) - \frac{1}{2}\|h_0\|_H^2 = \inf\{\ f(0,h) - \frac{1}{2}\|h\|_H^2;\ h \in H\ \}.$$

(3) For any h H with $h \neq 0$, $D^2 f(0,h_0)(h,h) - (h,h)_H < 0$.

Then for any $g \in \mathcal{G}(\mathcal{L};R)$, there exists a sequence $\{a_n\}_{n=0}^{\infty}$ such that

$$E[\ g(s,w(s)) \exp(\ \frac{1}{s} f(s,w(s)))\ ]$$

$$= \exp(\ -\frac{1}{2s}\|h_0\|_H^2 + \frac{1}{s} f(0,h_0)\ )\ (\ \sum_{n=1}^{N} a_n \cdot s^n\ + O(s^{N+1}))\ ,\ s \downarrow 0,$$

for any $N \geqq 0$. Moreover, as a formal power series in s

$$\sum_{n=1}^{\infty} a_n \cdot s^n$$

$$= \exp(f(o,h_0)) E[(\sum_{\alpha\in\mathfrak{A}} s^{\|\alpha\|/2} I^\alpha (\mathcal{B}^\alpha g(0,h_0))(1,w))$$

$$\times \{ 1 + \sum_{k\geq 1} \frac{1}{k!} (\sum_{\substack{\|\alpha\|\geq 3 \\ \alpha}} s^{\frac{|\alpha|}{2}-1} I^\alpha (\mathcal{B}^\alpha f(0,h_0)(1,w)))^k \}$$

$$\times \exp(\int_0^1 (D^2 f(0,h_0) w(\tau)) dw(\tau)))] .$$

Remark 4.4. It is easy to see that

$$a_0 = g(0,h_0) \exp(\mathcal{A} f(0,h_0)) \det_{(2)}(I_H - D^2 f(0,h_0))^{-1/2} ,$$

where $\det_{(2)}$ is the Carleman-Fredholm determinant. If $D^2 f(0,h_0)$ is a nuclear operator and if

$$\mathcal{A} f(0,h_0) = \frac{\partial}{\partial s} f(0,h_0) + \frac{1}{2} \operatorname{trace}_H D^2 f(0,h_0), \text{ then}$$

$$a_0 = g(0,h_0) \exp(\frac{\partial}{\partial s} f(0,h_0)) \det(I_H - D^2 f(0,h_0))^{-1/2}.$$

5. Splitting Wiener process and Bismut's expansion formula. Let K be a finite dimensional subspace of H. Let H_0 denote the orthogonal complement of K in H and B_0 denote the closure of H_0 in B. Let P_K denote the orthogonal projection onto K. Then for P-a.e.w, $P_K w(s)$, $s \geq 0$, is well defined. Set $w_0(s;w) = w(s) - P_K w(s)$, and let P_0 be the probability measure $W_0 = W_{B_0} = C([0,\infty);B_0)$ induced by P(dw) through $w_0(\cdot;w)$. Then (H_0,B_0,P_0) is also an abstract Wiener process. Thus we can define D, $\mathcal{L}$, $\mathcal{A}$, $\mathcal{G}(\mathcal{A};E)$ etc., for (H_0,B_0,P_0). To identify them from those for the original abstract Wiener process (H,B,P), we denote them by D_0, $\mathcal{L}_0$, $\mathcal{A}_0$, $\mathcal{G}(\mathcal{A}_0;E)$ etc.

For each $f \in \mathcal{G}(\mathcal{A};E)$, we define $\mathcal{R}f : [0,\infty)\times H_0 \times W_0 \to L^2_{loc}(K\to E,dk)$ by $\mathcal{R}f(s,h_0+w_0(s))(k) = f(s,h_0+w_0(s)+k)$. Let U_r , $r > 0$, denote $\{ k \in K ; \|k\|_H < r \}$. Then we have the following.

Theorem 5.1. For any $r > 0$ and integer n, if $f \in \mathcal{G}(\mathcal{A};E)$, then $\mathcal{R}f \in \mathcal{G}(\mathcal{A}_0;H_2^n(U_r;E))$. Moreover, if $f \in \mathcal{G}(\mathcal{A};E)$ is completely regular, then $\mathcal{R}f \in \mathcal{G}(\mathcal{A}_0;H_2^n(U_r;E))$ is also completely regular.

Now we can state Bismut's Expansion Theorem.

Theorem 5.2. Let $y \in R^N$. Suppose that $F \in \mathcal{G}(\mathcal{A};R^N)$ and is completely regular. Suppose moreover

(1) $\overline{\lim}_{s\downarrow 0}\ s \log E[\ \det\{(DF^i(s,w(s)),DF^j(s,w(s)))_{\mathcal{H}(R)}\}_{i,j=1,\dots,N}^{-p}] = 0$

for any $1 < p < \infty$,

(2) h_0 H is a unique minimum point of the function

$h \in \{\ h \in H\ ;\ F(0,h) = y\ \} \to \|h\|_H^2$,

(3) $T = DF(0,h_0) : H \to R^N$ has rank N, and

(4) the bilinear form V in ker T given by

$$V(k_1,k_2) = (\ k_1,k_2)_H + (\ T^*(T\cdot T^*)^{-1}D^2F(0,h_0)(k_1,k_2),\ h_0)_H\ ,$$

$k_1,k_2 \in \ker T$, is strictly positive definite.

Then the probability law of $F(s,w(s))$ on R^N under $P(dw)$ has smooth density $p(s,\cdot)$ and there exists $\{a_k\}_{k=0}^{\infty} \subset R$ such that

$$p(s,y) = (\tfrac{1}{2\pi s})^{N/2} \exp(\ -\ \tfrac{1}{2s}\|h_0\|_H^2)\ \{\ \sum_{k=0}^{N} a_k\cdot s^k + O(s^{N+1})\ \}\ ,\quad s \downarrow 0,$$

for any integer N. In particular

$$a_0 = \det(T\cdot T^*)^{-1/2}\det_{(2)}(V)^{-1/2}$$
$$\times \exp(\ (\ h_0,T^*(T\cdot T^*)^{-1}\{\ \mathcal{A}F(0,h_0) - \tfrac{1}{2}\operatorname{trace}_H P_K D^2F(0,h_0)\}\ )\ )$$
$$> 0\ ,$$

where $K = \text{Image } T^* = \text{Image } DF(0,h_0)^*$.

The sketch of the proof. Let $\psi_0 = F(0,h_0+\cdot) \in C^\infty(K;R^N)$. Then $\psi_0(0) = y$ and $\operatorname{grad}\psi_0(0) \neq 0$. Therefore there are $r > 0$ and $\varepsilon > 0$ and a smooth function $z : \{\ \psi \in C^2(\overline{U}_r;R^N);\ |\psi-\psi_0|_{C^2(\overline{U}_r)} < \varepsilon\ \} \to U_{r/4}$ such that $\psi(z(\psi)) = y$, $\psi(k) \neq y$ if $k \in \overline{U}_r\setminus\{z(\psi)\}$, and $|\operatorname{grad}\psi(k)| > \varepsilon$ if $k \in \overline{U}_r$. Let $\rho_i : H_2^{N+2}(U_r) \to R$, $i = 1,2$, be smooth functions and $0 < \delta_1 < \delta_2$ satisfying

$$\rho_1(\psi) = \begin{cases} 1 & \text{if } \|\psi-\psi_0\|_{H_2^{N+2}(\bar{U}_r;R^N)} < \delta_1 \\ 0 & \text{if } \|\psi-\psi_0\|_{H_2^{N+2}(\bar{U}_r;R^N)} > \delta_2 \end{cases} \quad \text{and}$$

$$\rho_2(\psi) = \begin{cases} 1 & \text{if } \|\psi-\psi_0\|_{H_2^{N+2}(\bar{U}_r;R^N)} < 2\delta_2 \\ 0 & \text{if } \|\psi-\psi_0\|_{C^2(\bar{U}_r;R^N)} > \varepsilon/2 \end{cases} .$$

Let $\phi \in C^\infty(K;R)$ satisfying $\phi(k) = \begin{cases} 1 & \text{if } |k| < \frac{r}{2} \\ 0 & \text{if } |k| > \frac{3}{4} r \end{cases}$.

Now split H to $H_0 \oplus K$, and let B_0, $w_0(s,w), \ldots$ as before. Then

$$\begin{aligned} & p(s,y) \\ = \; & E^P[\ \delta(\ F(s,w(s)) - y\ )\ ] \\ = \; & E^P[\ (\ 1 - \phi(P_K w(s)-h_0)\cdot\rho_1(\ F(s,h_0+w_0(s,w)))\ )\cdot\delta(F(s,w(s))-y)\ ] \\ & \quad + E^P[\ \phi(P_K w(s)-h_0)\cdot\rho_1(\ F(s,h_0+w_0(s,w)))\cdot\delta(F(s,w(s))-y)\ ] \\ = \; & I_1(s) + I_2(s)\ . \end{aligned}$$

Then for the second term $I_2(s)$, we have

$$\begin{aligned} & I_2(s) \\ = \; & E^P[\ \exp(\ -\frac{1}{2s}\|h_0\|_H^2 - \frac{1}{s}(w(s),h_0)_H\ )\cdot\phi(P_K w(s)-h_0) \\ & \qquad \times \rho_1(\ F(s,h_0+w_0(s,w)))\cdot\delta(F(s,h_0+w(s))-y)\ ] \\ = \; & (\frac{1}{2\pi s})^{N/2} \exp(\ -\frac{1}{2s}\|h_0\|_H^2\ ) \\ & \qquad \times E^{P_0}[\ \int_K dk\ \exp(\ -\frac{1}{2s}\|k\|_H^2 - \frac{1}{s}(k,h_0)_H\ )\ \phi(k) \\ & \qquad \qquad \times \rho_1(\ F(s,h_0+w_0(s)))\cdot\delta(\ F(s,h_0+w_0(s))(k)-y)\ ] \\ = \; & (\frac{1}{2\pi s})^{N/2} \exp(\ -\frac{1}{2s}\|h_0\|_H^2)\ E^{P_0}[\ g(s,w_0(s))\ \exp(\ \frac{1}{s} f(s,w_0(s)))], \end{aligned}$$

where $f(s,w_0(s)) = -(\ k(s,w_0(s)),\ h_0)_H - \frac{1}{2}|k(s,w_0(s))|_H^2$,

$$g(s,w_0(s)) = \phi(k(s,w_0(s)))\cdot\rho_1(\,F(s,h_0+w_0(s)))$$
$$\times \det_K(D_K\, F(s,h_0+w_0(s))(k(s,w_0(s)))$$
$$D_K\, F(s,h_0+w_0(s))(k(s,w_0(s)))^*)^{-1/2},$$

and $k(s,w_0(s)) = \rho_2(\,F(s,h_0+w_0(s)))\cdot z(\,F(s,h_0+w_0(s)))$.
Then from the assumption and Theorem 5.1, we have $k \in \mathcal{G}(\mathcal{A}_0;R^N)$ and $f,g \in \mathcal{G}(\mathcal{A}_0;R)$, and they are completely regular. Since $f(0,0) = 0$, $Df(0,0) = 0$ and $D^2f(0,0) = V$, Theorem 4.4. applies.

For the first term $I_1(s)$, by using Malliavin's integration by parts and Theorem 4.2., we have

$\overline{\lim_{s\downarrow 0}}\; s \log I_1(s) < -\frac{1}{2}\|h_0\|_H^2$. Thus we obtain our theorem.

Acknowledgements. The author acknowledges the hospitality at ZiF, University of Bielefeld, under Project No 2 (Mathematics and Physics). He is also grateful to Prof. J.M. Bismut and Prof. S. Watanabe for useful discussions.

References

[1] Azencott, R., Grandes déviations et applications, Cours de Probabilité de Saint Flour, Lecture Notes in Math. vol. 774, Springer, Berlin, 1978.

[2] Bismut, J.M., Large deviations and the Malliavin calculus, Progress in Math. vol. 45, Birkhäuser, Boston, Basel, Stuttgart, 1984.

[3] Gaveau, B., Principe de moindre action, propagation de la chaleur, estimées sous-elliptiques sur certains groupes nilpotents, Acta Math. 139 (1977), 96-153.

[4] Gross, L., Potential Theory on Hilbert Space, J. of Func. Analysis 1 (1967), 123-181.

[5] Kusuoka, S., and Stroock, D., Applications of the Malliavin calculus, Part I, Stochastic Analysis, Proc. of the Taniguchi International Symp. Katata and Kyoto 1982, ed. K. Ito, Kinokuniya, Tokyo, 1984.

[6] Malliavin, P. A., Calcul des variations stochastiques subordonné au processus de la chaleur, C. R. Acad. Sc. Paris Série I, t. 295 (1982), 167-172.

[7] Molchanov, S. A., Diffusion processes and Riemanian geometry, Russian Math. Surveys 30 (1975), 1-53.

AN ELEMENTARY APPROACH TO BROWNIAN MOTION ON MANIFOLDS

J.T. Lewis
Dublin Institute for Advanced Studies
10 Burlington Road
Dublin 4, Ireland

Talk given at the BiBoS - Symposium
"Stochastic Processes - Mathematics and Physics"
held at Zif - Bielefeld, September 1984.

§1. Introduction

In this talk I will describe an elementary approach to the study of Brownian motion on manifolds. It arose from reading the paper by Price and Williams [1] on Brownian motion on the unit sphere S^2 in $\mathbb{R}^3$. Their results were generalized to a hypersurface in $\mathbb{R}^d$ in [2] and to a submanifold of $\mathbb{R}^d$ of arbitrary co-dimension in [3]. This approach regards all the processes involved as processes on the ambient Euclidean space; it has the advantage that it lends itself to the martingale point of view; it has the disadvantage that all the objects of differential geometry which arise (covariant derivative, second fundamental form, Laplace-Beltrami operator,...) must be defined in an open neighbourhood of the submanifold. The casual reader is warned that there is already an extensive literature on Brownian motion on manifolds in which the differential geometry is treated from the intrinsic point of view; Ellworthy [4] is an excellent guide to this.

We begin by recalling the following equivalent definitions of BM($\mathbb{R}^d$), Brownian motion in $\mathbb{R}^d$:

1. A process B on $\mathbb{R}^d$ with $B_0=0$ is a BM($\mathbb{R}^d$) if and only if B_t is gaussian with $E[B_t]=0$ for each t and $E[B_s B_t^T] = (s\wedge t)1$ for each pair s,t. (Here the superscript T denotes 'transpose', and we regard $B_s B_t^T$ as a linear mapping on $\mathbb{R}^d$; the identity mapping on $\mathbb{R}^d$ is denoted by 1).

2. A process B on $\mathbb{R}^d$ with $B_0=0$ is a BM($\mathbb{R}^d$) if and only if B is a diffusion on $\mathbb{R}^d$ with generator $\frac{1}{2}\Delta$, where Δ is the Laplacian on $\mathbb{R}^d$.

3. A process B on $\mathbb{R}^d$ with $B_0=0$ is a BM($\mathbb{R}^d$) if and only if B is a semimartingale and

(i) $dB_t = dM_t$, where M is a continuous local martingale.

(ii) $d\langle BB^T\rangle_t = 1\,dt$. (1.1)

The equivalence of these three definitions is proved in Ikeda and Watanabe [5], for example. The first definition is the most elementary, but it cannot serve as a model for a definition of a Brownian motion on a manifold because the gaussian property will certainly not survive if the manifold is compact. The second definition is close to Einstein's original treatment and will serve as our model; we require simply that the process remains on the manifold for all t (almost surely), and that it is a diffusion whose generator is $\frac{1}{2}\Delta$, where Δ is now the Laplace-Beltrami operator. The third definition is a version of Lévy's martingale characterization of Brownian motion; it will serve as the model for our main result and it is the keystone of its proof.

§2. Submanifolds of Euclidean Space

We shall consider here submanifolds of $\mathbb{R}^d$ which are level sets of a C^2-function $f: U \to \mathbb{R}^r$ defined on an open set U in $\mathbb{R}^d$. We require that the level set $V = f^{-1}(c)$ be such that the derivative $f'(x)$ is of rank r for all x in V; then there is an open neighbourhood W of V such that $f'(y)$ has rank r for all y in W. The set W is made up of level sets of f, all having the same dimension. Let T_y be the kernel of $f'(y)$ for each y in W; then T_y is the tangent subspace at y to the unique level set of f through the point y, and we denote by $P(y)$ the orthogonal projection of $\mathbb{R}^d$ onto T_y. The orthogonal complement $T_y^{\perp}$ of T_y is the normal subspace at y, and we denote by $\overset{\perp}{P}(y)$ the orthogonal projection of $\mathbb{R}^d$ onto $T_y^{\perp}$. We will say that a vector field $v: W \to \mathbb{R}^d$ is a tangent vector field if $v(y)$ lies in T_y for each y in W; and we will say that it is a normal vector field if $v(y)$ lies in $\overset{\perp}{T}(y)$ for each y in W. Given a pair v,w of tangent vector fields v,w defined on W we decompose the derivative $(v\cdot\text{grad } w)(y)$ of w in the direction of v as

$$(v\cdot\text{grad } w)(y) = (\nabla_v w)(y) + s_y(v,w) \tag{2.1}$$

where

$$(\nabla_v w)(y) = P(y)\,(v\cdot\text{grad } w)(y), \tag{2.2}$$

and

$$s_y(v,w) = \overset{\perp}{P}(y)\,(v\cdot\text{grad } w)(y). \tag{2.3}$$

When restricted to V, the tangent vector field $\nabla_v w$ is called the covariant-derivative of w with respect to v, and the normal vector field $s(v,w)$ is called the second fundamental form of the imbedding of V in $\mathbb{R}^d$. We define another normal vector field j on W by

$$j(y) = \tfrac{1}{2}\, \mathrm{trace}_{T_y}(s_y). \tag{2.4}$$

Let $\{n_1,\ldots,n_r\}$ be an orthonormal family of normal vector fields on W; the set $\{n_1(y),\ldots,n_r(y)\}$ is an orthonormal basis for $T_y^{\perp}$. (We can construct such a family by taking the components $\{f^1,\ldots,f^r\}$ of f with respect to an orthonormal basis for $\mathbb{R}^r$ and applying the Gram-Schmidt process to $\{\mathrm{grad}\ f^1,\ldots,\ \mathrm{grad}\ f^r\}$). Then

$$\begin{aligned} S_y(v,w) &= \sum_{j=1}^{r} n_j(y)\{(v\cdot \mathrm{grad}\ w)(y)\cdot n_j(y)\} \\ &= -\sum_{j=1}^{r} n_j(y)(v(y)\cdot n_j'(y)w(y)), \end{aligned} \tag{2.5}$$

since $n_j(y)\cdot w(y) = 0$ for $j=1,\ldots,r$. Thus

$$j(y) = -\tfrac{1}{2}\sum_{j=1}^{r} n_j(y)\ \mathrm{trace}\ [P(y)n_j'(y)] \tag{2.6}$$

$$= -\tfrac{1}{2}\sum_{j=1}^{r} n_j(y)\{(\mathrm{div}\ n_j) - \sum_{k\neq j}(n_k(y)\cdot n_j'(y)n_k(y))\}. \tag{2.7}$$

In the case of a hypersurface $(r=1)$ the expression (2.7) can be written

$$j(y) = \frac{d-1}{2} H(y)n(y), \tag{2.8}$$

where $H(y)$ is the mean curvature at y of the level surface through y and n is the orienting vector field, while (2.7) yields the computationally useful formula

$$j(y) = -\tfrac{1}{2}\, n(y)(\mathrm{div}\ n)(y). \tag{2.9}$$

The covariant derivative $(\nabla g)(y)$ of a function $g: W \to \mathbb{R}$ is defined by

$$(\nabla g)(y) = P(y)(\mathrm{grad}\ g)(y), \tag{2.10}$$

and the Laplace-Beltrami operator Δ by

$$(\Delta g)(y) = \mathrm{trace}\ ((\nabla^2 g)(y)). \tag{2.11}$$

It follows from (2.10) that

$$(\Delta g)(y) = \text{trace}\,[P(y)\{\text{grad}(P(y)\,\text{grad}\, g(y))\}] = \text{trace}\,[P(y)g''(y)]$$
$$= -\sum_{j=1}^{r} (\text{grad}\, g(y)\cdot n_j(y))\ \text{trace}\,[P(y)n_j'(y)]. \tag{2.12}$$

Rewriting (2.12), using (2.6), we have

$$\tfrac{1}{2}(\Delta g)(y) = \tfrac{1}{2}\,\text{trace}\,[P(y)g''(y)] - j(y)\cdot(\text{grad}\, g)(y), \tag{2.13}$$

an identity which will prove useful in the next section.

§3. Brownian Motion on a Submanifold

Let $V = f^{-1}(c)$ be as described in §2. We claim that a process X in $\mathbb{R}^d$, with $f(X_0) = c$ and

$$dX_t - j(X_t)dt = P(X_t)dB_t, \tag{3.1}$$

is a BM(V), a Brownian motion on the submanifold $V = f^{-1}(c)$; here B is a BM($\mathbb{R}^d$), a Brownian motion on $\mathbb{R}^d$.

Now X is a diffusion, since it satisfies an Itô equation; we have to show that its generator is $\frac{1}{2}\Delta$ and that it remains on the surface for $t>0$. Let g be an arbitrary C^2-function $g : W \to \mathbb{R}$, and apply Itô's formula to the process $g(X)$:

$$dg(X_t) = (\text{grad}\, g)(X_t)\cdot dX_t + \tfrac{1}{2}\,\text{trace}\,[g''(X_t)d\langle XX^T\rangle_t]. \tag{3.2}$$

From (3.1) and (1.1) we have

$$d\langle XX^T\rangle_t = P(X_t)dt, \tag{3.3}$$

so that

$$dg(X_t) = dN_t + j(X_t)\cdot(\text{grad}\, g)(X_t)dt + \tfrac{1}{2}\,\text{trace}\,[g''(X_t)P(X_t)]dt \tag{3.4}$$

where

$$dN_t = (\text{grad}\, g)(X_t)\cdot P(X_t)dB_t. \tag{3.5}$$

Thus

$$dg(X_t) - \tfrac{1}{2}(\Delta g)(X_t)dt = dN_t, \tag{3.5}$$

where N_t is a continuous local martingale; we conclude that $\frac{1}{2}\Delta$ is the generator of the diffusion X. It remains to show that X remains on $V = f^{-1}(c)$ for $t>0$. Let $g = f^j, j=1,\dots,r$; then

$$dN_t = (\text{grad } f^j)(X_t)\cdot P(X_t)dB_t = 0$$

since $(\text{grad } f^j)(y)$ is orthogonal to T_y, and

$$(\nabla f^j)(y) = P(y)(\text{grad } f^j)(y) = 0,$$

for the same reason. It follows from (3.4) that $df^j(X_t) = 0$ for $j=1,\dots,r$. Thus X stays on $V = f^{-1}(c)$ for $t>0$ since it starts there.

Remark: The equation (3.1) for Brownian motion on a submanifold of Euclidean space was given by Baxendale [6].

§4. Martingale Characterization

The description of Brownian motion on $V = f^{-1}(c)$ given in §3 suggests the following

Martingale Characterization of BM(V):

A process X on $\mathbb{R}^d$ with $f(X_0) = c$ is a BM(V) if and only if X is a semimartingale such that

(1) $dX_t - j(X_t)dt = dM_t$, *where M is a continuous local martingale.*

(2) $d\langle XX^T\rangle_t = P(X_t)dt$.

We have to show that, given a semimartingale on $\mathbb{R}^d$ satisfying (1) and (2), there exists B, a BM($\mathbb{R}^d$), such that

$$dM_t = P(X_t)dB_t. \tag{4.1}$$

Let $\tilde{B}$ be a BM($\mathbb{R}^d$) which is independent of X, so that

$$d\langle\tilde{B}\tilde{B}^T\rangle_t = 1\, dt,\ d\langle X\tilde{B}^T\rangle = 0, \tag{4.2}$$

and let $\mathcal{B}$ be a process on $\mathbb{R}^d$ such that $B_0=0$ and

$$d\mathcal{B}_t = P(X_t)dX_t + P^{\perp}(X_t)d\tilde{B}_t; \tag{4.3}$$

then by (2) and (4.2) we have

$$d\langle\tilde{B}\tilde{B}^T\rangle = P(X_t)dt + P^{\perp}(X_t)dt = 1\ dt. \tag{4.4}$$

It follows, by the martingale characterization of BM($\mathbb{R}^d$), that $\tilde{B}$ is a BM($\mathbb{R}^d$) and, by (1), that

$$P(X_t)dM_t = P(X_t)dB_t. \tag{4.5}$$

It remains to show that $P(X_t)dM_t = dM_t$. Consider the process N on $\mathbb{R}^d$ such that $N_0=0$ and

$$dN_t = P^{\perp}(X_t)dM_t. \tag{4.6}$$

Then

$$d\langle NN^T\rangle_t = P^{\perp}(X_t)P(X_t)P^{\perp}(X_t)dt = 0, \tag{4.7}$$

so that NN^T is also a continuous local martingale; but NN^T is non-negative so that NN^T is constant almost surely, and so $dN_t=0$ and

$$dM_t = P(X_t)dM_t = P(X_t)dB_t. \tag{4.8}$$

§5. Examples

(1) Hypersurfaces in $\mathbb{R}^d$

In this case, r=1 and

$$j(x) = \frac{d-1}{2}H(x)n(x), \tag{5.1}$$

where H(x) is the mean curvature of V at x, and n is the orienting normal vector field. Then _a_ BM(V) _is a martingale in the ambient Euclidean space if and only if the mean curvature of V vanishes identically._ (Compare [7]).

It follows from (3.1) that, if X is such that

$$dX_t - \frac{(d-1)}{2}H(X_t)n(X_t)dt = P(X_t)dB_t, \tag{5.2}$$

then X is a BM(V). It follows from the martingale characterization that an alternative equation for BM(V) is

$$dX_t - \frac{(d-1)}{2}H(X_t)n(X_t)dt = d\tilde{B}_t n(X_t), \tag{5.3}$$

where $\tilde{B}$ is a BM(so(d)), a Brownian motion in the Lie algebra of the orthogonal group SO(d), since $d\langle XX^T\rangle_t = P(X_t)dt$; see [2].

(2) <u>The unit sphere S^2 in $\mathbb{R}^2$</u>

In the special case of S^2, the unit sphere in $\mathbb{R}^3$, we take $n(x)=x$, the outward normal at x; then the principal curvatures are both equal to -1, so that $j(x)=-x$. The projection $P(x)$ onto the tangent space at x is given by $P(x) = (1-xx^T)$. Then (5.2) yields the equation of Stroock[7]

$$dX_t + X_t dt = (1 - X_t X_t^T)dB_t. \tag{5.4}$$

On the other hand, (5.3) yields the equation of Price and Williams [1]:

$$dX_t + X_t dt = X_t \times dB_t. \tag{5.5}$$

(3) <u>Curves in $\mathbb{R}^d$</u>

Let $s \to x(s)$ be a C^2-curve in $\mathbb{R}^d$, parametrized by arc length; then the tangent vector $t(s)$ at $x(s)$ is given by

$$t(s) = \frac{dx}{ds}(s) \tag{5.6}$$

and

$$\frac{dt}{ds}(s) = k(s)n(s) \tag{5.7}$$

where $n(s)$ is the principal normal at $x(s)$ and $k(s)$ is the curvature. Then

$$j(x(s)) = \tfrac{1}{2}k(s)n(s), \tag{5.8}$$

and

$$P(x(s)) = t(s)t(s)^T. \tag{5.9}$$

Now let b be a BM($\mathbb{R}^1$) and put $X_t = x(b_t)$. Then X is a process in $\mathbb{R}^d$ beginning at $x(0)$ and

$$dX_t = \frac{dx}{ds}(b_t)db_t + \tfrac{1}{2}\frac{d^2}{ds^2}x(b_t)dt = t(b_t)db_t + \tfrac{1}{2}k(b_t)n(b_t)dt, \tag{5.10}$$

so that

$$dX_t - j(X_t)dt = t(b_t)db_t. \tag{5.11}$$

It follows from (5.11) that

$$d\langle XX^T\rangle_t = t(b_t)t(b_t)^T dt; \tag{5.12}$$

using (5.8) we have

$$d\langle XX^T\rangle_t = P(X_t)dt.$$

By the martingale characterization, it follows that X is a Brownian motion on the curve $s \to x(s)$.

§6. Martingale Representation

Let X be a Brownian motion on $V = f^{-1}(c)$ starting at x, and let Y be defined by $Y_0=0$ and

$$dY_t = P(X_t)dX_t, \tag{6.1}$$

so that dY_t is the tangential component of dX_t. Let $\tilde{X}$ be another Brownian motion on $V = f^{-1}(c)$ starting at x, and let $\tilde{Y}$ be defined by $Y_0=0$ and

$$d\tilde{Y}_t = P(\tilde{X}_t)d\tilde{X}_t. \tag{6.2}$$

Suppose that $\tilde{X}$ is adapted to the filtration of X; then we have the following

Martingale Representation: The processes Y and $\tilde{Y}$ are related by the Itô equation

$$d\tilde{Y}_t = C_t dY_t \tag{6.3}$$

where

(1) for each t, C_t is an orthogonal transformation such that

$$C_t n(X_t) = n(\tilde{X}_t) \tag{6.4}$$

for each unit normal vector field n on V.

(2) the process C is X-predictable.

Let $\{n_1,\ldots,n_r\}$ be an orthonormal set of normal vector fields on V; let $\{b^1,\ldots,b^r\}$ be a set of independent $BM(\mathbb{R}^1)$ - processes independen of both X and $\tilde{X}$ so that

$$d\langle X^i b^j\rangle = d\langle \tilde{X}^i b^j\rangle = 0, \ i,j=1,\ldots,r, \tag{6.5}$$

and

$$d\langle b^i b^j\rangle = \delta_{ij}dt. \tag{6.6}$$

Then, by the argument in §4, the processes B and $\tilde{B}$ such that $B_0=\tilde{B}_0=0$ and

$$dB_t = dY_t + \sum_{j=1}^{r} n_j(X_t)db^j, \quad d\tilde{B}_t = d\tilde{Y}_t + \sum_{j=1}^{r} n_j(\tilde{X}_t)db^j \tag{6.7}$$

are both BM($\mathbb{R}^d$) and X and $\tilde{X}$ satisfy

$$dX_t - j(X_t)dt = P(X_t)dB_t, \quad d\tilde{X}_t - j(\tilde{X}_t)dt = P(\tilde{X}_t)d\tilde{B}_t. \tag{6.8}$$

Moreover, $\tilde{B}$ is B-predictable so that, by the martingale representation theorem for BM($\mathbb{R}^d$), there exists a B-predictable process C of orthogonal transformations on $\mathbb{R}^d$ such that

$$d\tilde{B}_t = C_t dB_t. \tag{6.9}$$

Hence, from (6.7), we have

$$C_t dY_t + \sum_{j=1}^{r} C_t n_j(X_t)db^j = d\tilde{Y}_t + \sum_{j=1}^{r} n_j(\tilde{X}_t)db^j; \tag{6.10}$$

Forming the bracket process of both sides with the process b^k, using (6.5) and (6.6), we have

$$C_t n_k(X_t)dt = n_k(\tilde{X}_t)dt, \tag{6.11}$$

establishing (6.4), and (6.3) follows by subtraction. It follows from (6.4) that C can be chosen to be X-predictable.

Special Cases:

(1) For a hypersurface ($r=1$); taking n to be the orienting vector field, the map $x \to n(x)$ is the Gauss map.

(2) Specializing to S^2, the unit sphere in $\mathbb{R}^3$, we have $n(x)=x$ and we recover the result of Price and Williams [1]:

Let X and $\tilde{X}$ be BM(S^2) - processes starting at x; suppose that $\tilde{X}$ is adapted to the filtration of X. Then the tangential increments dY and $d\tilde{Y}$ are related by the Itô equation

$$d\tilde{Y}_t = C_t dY_t \tag{6.12}$$

where (1) for each t, C_t is an orthogonal transformation such that

$$C_t X_t = \tilde{X}_t, \tag{6.13}$$

(2) the process C is X-predictable.

Acknowledgements: It is a pleasure to thank Michiel van den Berg and Paul McGill for many stimulating discussions.

References

[1] G.C. Price, D. Williams: Rolling with 'Slipping': I. Sém. Prob. Paris XVII. Lect. Notes in Maths. 986, Berlin-Heidelberg-New York: Springer 1983.

[2] M. van den Berg, J.T. Lewis: Brownian Motion on a Hypersurface, Bull. London Math. Soc. (in press).

[3] J.T. Lewis: Brownian Motion on a Submanifold of Euclidean Space, (preprint: DIAS-STP-84-48).

[4] K.D. Ellworthy: Stochastic Differential Equations on Manifolds, LMS Lecture Notes 70, Cambridge: C.U.P. 1981.

[5] N. Ikeda, S. Watanabe: Stochastic Differential Equations and Diffusion Processes. Amsterdam-Oxford-New York: North Holland 1981.

[6] P.H. Baxendale: Wiener Processes on Manifolds of Maps, Proc. Royal Soc. Edinburgh 87A (1980) 127-152.

[7] R.W.R. Darling: A Martingale on the Imbedded Torus, Bull. London Math. Soc. 15, 221-225 (1983).

THE STOCHASTIC MECHANICS OF THE GROUND-STATE OF THE HYDROGEN ATOM

A. Truman
Department of Mathematics
University College of Swansea
Singleton Park, Swansea SA2 8PP
Wales

J.T. Lewis
Dublin Institute for Advanced Studies
10 Burlington Road, Dublin 4
Ireland

Talk given by A. Truman at the BiBoS - Symposium

Bielefeld, September, 1984.

§1. Introduction

The prospect of a formulation of quantum mechanics on path space, offered by Nelson's stochastic mechanics, is attractive to many. So far, it has been shown to be in complete agreement with the Schrödinger theory at the level of particle densities and spectra. But stochastic mechanics, because it operates at the sample-path level, seems to contain more information than the Schrödinger theory. The sample-paths in stochastic mechanics satisfy Nelson's generalization of Newton's equation (see §3 below); it is tempting, therefore, to conjecture that the sample-paths of the process provide us with the ensemble of actual particle paths. In this paper we explore one consequence of this conjecture.

If the Nelson theory is to be more than a very elegant reformulation of Schrödinger's wave-mechanics, one must find areas where the predictions of stochastic mechanics and quantum theory differ. One area which merits investigation is the calculation of first hitting times in stochastic mechanics and first detection times in quantum theory. As a first step in this direction we consider the stochastic mechanics of the stationary states of the Hydrogen atom, a readily identifiable quantum system.

In this paper we analyze in detail the diffusion process corresponding to the ground state of the Hydrogen atom; we obtain a skew-product formula for the process and give detailed results on first-hitting times. In [1] we describe how first arrival times of quantum particles can be calculated in the quantum theory of counting processes due to Davies [2]. Finally, we indicate how our results may be extended to excited states.

Since this is addressed to a mixed audience of physicists and mathematicians, we have concentrated on explaining the ideas. The proofs are given in outline; details will be given in [1].

It is a pleasure to thank David Williams and Barry Simon for helpful conversations and to acknowledge SERC support through research grant GR/C/13644.

§2. Diffusion Processes and Schrödinger's Equation

First we recall some results about diffusion processes (see [3] for details). Let X be a process on $\mathbb{R}^d$ satisfying the Itô equation

$$dX_t = b(X_t,t)dt + dB_t \tag{2.1}$$

where B is a BM($\mathbb{R}^d$), a Brownian motion on $\mathbb{R}^d$, so that each component B^j of B is a Gaussian process with $E[B^j_t]=0$ and

$$E[B^j_s B^k_t] = \delta_{jk}\cdot s\wedge t,\ i,j = 1,\dots,d. \tag{2.2}$$

Here E[·] denotes the expectation with respect to Wiener measure P. If the drift b is sufficiently well-behaved, the process X has a transition density p(x,s ; y,t), defined for t>s by

$$P[X_t \in A,\ X_s = x] = \int_A p(x,s\,;\,y,t)dy, \tag{2.3}$$

and p is the fundamental solution of the forward Kolmogorov equation

$$\frac{\partial p}{\partial t} = \operatorname{div}_y\left(\tfrac{1}{2}\operatorname{grad}_y p - b(y,t)p\right) \overset{\text{def}}{=} L_y p, \tag{2.4}$$

subject to the condition

$$p(x,t\,;\,y,t) = \delta(x-y). \tag{2.5}$$

The subscript y on the differential operators indicates that differentiation is with respect to the final point y. Regarded as a function of the starting point x, the transition density p satisfies

$$\frac{\partial p}{\partial t} = \tfrac{1}{2}\Delta_x p + b(x,t)\operatorname{grad}_x p \overset{\text{def}}{=} L^* p, \tag{2.6}$$

the backward Kolmogorov equation. Here L^* is the formal L^2 - adjoint of L.

Now let f(x,t) be a classical solution of the Schrödinger equation

$$i\frac{\partial f}{\partial t} = -\tfrac{1}{2}\Delta_x f + Vf, \tag{2.7}$$

where V is a real-valued potential function. Multiplying by f*, the complex conjugate of f, and equating real parts, we find that the continuity equation

$$\frac{\partial \rho}{\partial t} + \operatorname{div} j = 0 \tag{2.8}$$

is satisfied by the quantum mechanical probability density $\rho = |f|^2$ and the quantum mechanical probability current $j = -\frac{i}{2}(f^* \operatorname{grad} f - f \operatorname{grad} f^*)$. Now assume that f is nowhere zero and write $f = \exp(R + iS)$, where R and S are real-valued; then

$$\rho = e^{2R}, \; j = e^{2R} \operatorname{grad} S. \tag{2.9}$$

Hence we see that (2.8) may be written as

$$\frac{\partial \rho}{\partial t} = \operatorname{div}\left(\tfrac{1}{2} \operatorname{grad} \rho - \rho \operatorname{grad}(R + S)\right). \tag{2.10}$$

Comparing this equation with (2.4), we recognize it as the forward Kolmogorov equation with drift

$$b = \operatorname{grad}(\operatorname{Re} \log f + \operatorname{Im} \log f). \tag{2.11}$$

This leads to:

Proposition 0

Let V be a real-valued potential function and let f be a classical solution of the Schrödinger equation

$$i\frac{\partial f}{\partial t} = -\tfrac{1}{2}\Delta_x f + Vf. \tag{2.12}$$

Suppose that f is nowhere zero and that the corresponding drift b is a continuous function of (x,t) satisfying

$$|b(x,t)| < M(|x| + 1), \tag{2.13}$$

for some positive constant M. Then, if there exist positive constants B and C such that

$$|f(x,t)| < B \exp(C|x|^2), \tag{2.14}$$

and if the transition density p(x,0 ; y,t) exists for the diffusion process X which satisfies

$$dX_t = b(X_t,t)dt + dB_t, \tag{2.15}$$

then

$$|f(y,t)|^2 = \int_{\mathbb{R}^d} p(x,0\,;\,y,t)\,|f(x,0)|^2 dx. \tag{2.16}$$

The proof is a straightforward application of the uniqueness theorem for parabolic equations (see [4]). As an application we cite the following

Example:

Consider the solution

$$f(x,t) = \exp\{-\tfrac{1}{2}(x-ae^{it})^2 - \frac{a^2}{4}(1-e^{-2it}) - \frac{it}{2}\} \tag{2.17}$$

of the equation

$$i\frac{\partial f}{\partial t} = -\tfrac{1}{2}\frac{\partial^2 f}{\partial x^2} + \tfrac{1}{2}x^2 f. \tag{2.18}$$

Then

$$|f(x,t)|^2 = \exp\{-(x-a\cos t)^2\}. \tag{2.19}$$

The corresponding diffusion process X satisfies

$$dX_t = -(X_t - a\cos t + a\sin t)dt + dB_t, \tag{2.20}$$

and the transition density p is given explicitly by

$$p(x,0\,;\,y,t) = (2\pi t)^{-\frac{1}{2}}\exp\{-\frac{(y-x)^2}{2t} + A(t,y) - A(0,x)\}E(x,0\,;\,y,t) \tag{2.21}$$

where

$$E(x,0\,;\,y,t) = E[\exp\{t\int_0^1 F(ut,(y-x)u + t^{\frac{1}{2}}b(u))du\}]; \tag{2.22}$$

here b is the Brownian bridge with $b(0) = b(1) = 0$, and

$$A(u,x) = -\tfrac{1}{2}(x-a\cos u + a\sin u)^2, \tag{2.23}$$

$$F(u,x) = A(u,x) + x(a\cos u + a\sin u) + \tfrac{1}{2}. \tag{2.24}$$

As Nelson [5] has pointed out, not all diffusions arise in this way; those which do are the conservative diffusions. To describe them, we need some definitions: the mean forward derivative D_+ is defined by

$$D_+X_t = \lim_{h\downarrow 0}\frac{1}{h}E[X_{t+h} - X_t\,|\,X_t]. \tag{2.25}$$

From (2.15) we deduce that

$$D_+X_t = b(X_t,t) = (\mathrm{grad}\,R)(X_t,t) + (\mathrm{grad}\,S)(X_t,t). \tag{2.26}$$

The mean backward derivative D_- is defined by

$$D_-X_t = \lim_{h\downarrow 0}\frac{1}{h}E[X_t - X_{t-h}\,|X_t]; \tag{2.27}$$

Nelson showed that

$$D_-X_t = b(X_t,t) - \text{grad }(\log \rho)(X_t,t) \tag{2.28}$$

so that we have

$$D_-X_t = (\text{grad } S)(X_t,t) - (\text{grad } R)(X_t,t). \tag{2.29}$$

This leads us to define the backward drift b_* as

$$b_*(x,t) = (\text{grad } S)(x,t) - (\text{grad } R)(x,t). \tag{2.30}$$

The kinetic energy T of the diffusion process is defined as

$$T(X_t,t) = \tfrac{1}{4}\{b(X_t,t)^2 + b_*(X_t,t)^2\} = \tfrac{1}{2}\{|\text{grad } S|^2(X_t,t) + |\text{grad } R|^2(X_t,t)\}. \tag{2.31}$$

A diffusion is said to be conservative if

$$\frac{d}{dt}E[T(X_t,t) + V(X_t)] = 0. \tag{2.32}$$

It is straightforward to check that the diffusion in the above example is conservative.

We turn now to diffusions associated with stationary states; we begin with an elementary proposition:

Proposition 1

Let $|f_E|$ be a classical solution of

$$-\tfrac{1}{2}\Delta_x|f_E| + V|f_E| = E|f_E| \tag{2.33}$$

such that $|f_E|>0$, where V is a real-valued potential which is bounded below. Let $g_E(x,y\,;t) = p(x,0\,;y,t)$ be the transition density for the diffusion process corresponding to the stationary state $e^{-iEt}|f_E|(x)$; then

$$g_E(x,y\,;t) = |f_E|(y)E[\exp\{\textstyle\int_0^t(E-V(x+B_s))ds\}\,|x+B_t = y]\,|f_E|(x)^{-1}, \tag{2.34}$$

where B is a $BM(R^d)$ with $B_0=0$. That is,

$$g_E(x,y\,;t) = |f_E|(y)\exp\{-t(H-E)\}(x,y)\,|f_E|(x)^{-1}, \tag{2.35}$$

where $H = -\tfrac{1}{2}\Delta + V$ is the Hamiltonian operator.

The proof is a simple application of the Girsanov-Cameron-Martin formula and the Feynman-Kac formula.

Remarks: (1) The process M given by $M_t = \exp\{\int_0^t(E-V(x+B_s))ds\}$ is a martingale with respect to the filtration of the Brownian motion B.

(2) The formula in (2.34) is valid whenever the Feynman-Kac formula holds

(3) The proposition enables us to discuss the asymptotics of $g_E(x,y;\, t)$:

(a) The large-time asymptotics can be summarized by

$$\int_{\mathbb{R}^d} g_E(x,y;\, t)h(x)dx \sim c|f_E|^2(y) \text{ as } t \sim \infty,$$

where $c = \int_{\mathbb{R}^d} h(x)dx$; this is merely an ergodicity result.

(b) Let $F_E^k(x,t) = \exp(-iEt/k)|f_E^k|(x)$, where $|f_E^k|$ satisfies $-\frac{k^2}{2}\Delta|f_E^k| + V|f_E^k| = E|f_E^k|$; the asymptotics for small k can be deduced for the diffusion process associated with this stationary state. For example, assume that V has a unique minimum at x=a and satisfies some mild subsidiary conditions. Using the results of Simon [6] or Davies and Truman [7], we can then deduce that

$$\lim_{k \downarrow 0} \{ -k \log g_E(x,y;\, t)\} = \rho(a,y) - \rho(a,x) + A(x,y;\, t), \text{ where}$$

$$A(x,y;\, t) = \inf_{c \in X_0^t} \{ \tfrac{1}{2}\int_0^t |\dot{c}|^2(s)ds - \int_0^t V(c(s))ds\},\ X_0^t = \{ c : [0.t] \to \mathbb{R}^d$$

absolutely continuous, $c(0) = x$, $c(t) = y\}$ and $\rho(x,y) = \inf_t A(x,y;\, t)$ is the Agmon metric.

(4) The proposition reminds us of the extremely useful result that, for non-zero stationary states $|f_E|$ and self-adjoint quantum mechanical Hamiltonians $H = -\frac{1}{2}\Delta + V$, the generator L of the associated diffusion satisfies

$$L = -|f_E|(H-E)|f_E|^{-1},\ L^* = -|f_E|^{-1}(H-E)|f_E|.$$

We shall make use of this in the next section

§3. Stochastic Mechanics of the Ground State of the Hydrogen Atom

In this section, we reinstate the reduced particle mass m and Planck's constant $\hbar = h/2\pi$; thus we will be concerned with the Schrödinger equation

$$i\hbar\frac{\partial f}{\partial t} = -\frac{\hbar^2}{2m}\Delta f + Vf. \qquad (3.1)$$

The associated diffusion X satisfies

$$dX_t = b(X_t,t)dt + (\frac{\hbar}{m})^{\frac{1}{2}}dB_t, \qquad (3.2)$$

provided f(x,t) is nowhere zero, where b is given by (2.11). The conventions of §2 remain in force, and d=3.

We shall be interested particularly in the diffusion associated with the ground state f_E of the Hydrogen atom with nucleus of charge Ze; in Gaussian units we have $V(x) = -Ze^2/|x|$ and

$$f_E(x,t) = N \exp(-|x|/a) \exp(-iEt/\hbar), \tag{3.3}$$

where a is Bohr radius and E is the ground state energy. (In Gaussian units, we have $a = \hbar^2/me^2Z$ and $E = -\hbar^2/2ma^2$.) The ground state process X satisfies

$$dX_t = -\frac{\hbar}{ma}\frac{X_t}{|X_t|}dt + \left(\frac{\hbar}{m}\right)^{\frac{1}{2}} dB_t. \tag{3.4}$$

The forward and backward derivatives are used to define a mean acceleration $\frac{1}{2}(D_+D_-+D_-D_+)X$ of a diffusion process X; for the ground-state process the mean acceleration satisfies Nelson's generalization of Newton's equation:

$$\frac{m}{2}(D_+D_-+D_-D_+)X_t = -Ze^2\frac{X_t}{|X_t|^3}. \tag{3.5}$$

To see this, use the formulae (which come from (3.4) and Itô's formula)

$$D_\pm X_t = \mp\frac{\hbar}{ma}\frac{X_t}{|X_t|}, \tag{3.6}$$

$$D_\pm h(X_t) = \text{grad } h\cdot D_\pm X_t \pm \frac{\hbar}{2m}(\Delta h)(X_t), \tag{3.7}$$

for an arbitrary smooth function h. When $h(x) = x/|x|$, we have grad $h\cdot D_\pm X=0$ and $(\Delta h)(x) = -2x/|x|^3$, so that $D_+D_-X_t = D_-D_+X_t = -\frac{\hbar^2}{m^2a}\frac{X_t}{|X_t|^3}$.

Nelson has shown that his generalization of Newton's equation is valid under very wide conditions. This leads to the suggestion that the sample paths could be thought of as the unseen trajectories of the quantum mechanical particle; we now examine the sample paths of the ground-state process in more detail. We begin with:

Proposition 2

The process $|X|$ satisfies

$$d|X_t| = \left(\frac{\hbar}{m}\frac{1}{|X_t|} - \frac{\hbar}{ma}\right)dt + \left(\frac{\hbar}{m}\right)^{\frac{1}{2}} db_t, \tag{3.8}$$

where b *is a* $BM(\mathbb{R}^1)$ *and* $|X_0|$ *has a distribution proportional to*

$r^2 \exp(-2r/a)$. The process $\hat{X} = \frac{X}{|X|}$ is given by $\hat{X}_t = Z(\frac{\hbar}{m}\int_0^t \frac{ds}{|X_s|^2})$, where Z is a $BM(S^2)$, a Brownian motion on S^2, the unit sphere in $\mathbb{R}^3$, with initial distribution uniform on S^2.

Proof: Using (3.4) and Itô's formula, we get (3.8) with $db = \hat{X}\cdot dB$; by the martingale characterization of Brownian motion, b is a $BM(\mathbb{R}^1)$ since $d\langle b\rangle_t = \hat{X}_t\cdot\hat{X}_t dt = dt$. Now apply Itô's formula to $\hat{X} = \frac{X}{|X|}$; we have

$$d\hat{X} = \left(\frac{\hbar}{m}\right)^{\frac{1}{2}} \frac{1}{|X|}(1 - \hat{X}\hat{X}^T)dB - \frac{\hbar}{m}\frac{\hat{X}}{|X|^2}dt. \tag{3.9}$$

Let $u(t)$ be the inverse of $t \to \frac{\hbar}{m}\int_0^t \frac{ds}{|X_s|^2}$, and put $Z_t = \hat{X}_{u(t)}$; then, by §2.5 of McKean [9], we have

$$dZ_t + Z_t dt = (1 - ZZ^T)dB. \tag{3.10}$$

We recognize (3.10) as Stroock's equation for $BM(S^2)$ (see the contributio by J.T. Lewis in this volume) and the result follows.

We study the radial process $|X|$ in more detail:

Proposition 3

Let $g_r(x,y;\ t)dy = P[\,|X_t| \in dy,\ |X_0| = x]$ for x,y in $(0, \infty)$. Let H_r be the radial part of the Hydrogen atom Hamiltonian;

$$H_r = -\frac{\hbar^2}{2m}\cdot\frac{1}{x^2}\frac{d}{dx}\,x^2\frac{d}{dx} - Ze^2\frac{1}{x}, \tag{3.11}$$

and let $f_0(x) = \exp(-x/a)$. Then

$$g_r(x,y;\ t) = f_0(y)\exp\{-t\,(H_r - E)/\hbar\}(x,y)f_0(x)^{-1} \tag{3.12}$$

and, for each $t>0$, we have

$$-\int_0^\infty \frac{\partial}{\partial t}\{\exp(-tE/\hbar)g_r(x,x;\ t)\}\,dx = \sum_{n=0}^\infty \frac{E_n}{\hbar}\exp(-tE_n/\hbar), \tag{3.13}$$

where $E_n (n = 0,1,\ldots)$ are the eigenvalues of H_r.

The proof is a straightforward calculation. Next we investigate first hitting times for the radial process.

§4. First Hitting Times for the Radial Process

In this section we present some results on the distribution of hitting times for the radial process associated with the ground-state of the Hydrogen atom.

Proposition 4

Let S_r be a sphere of radius r, with centre at the nucleus of the Hydrogen atom, and assume that the initial distribution of the radial ground-state process $|X|$ is the quantum mechanical ground-state probabili distribution, proportional to $x^2\exp(-2x/a)$. Let $T(c,b)$ be the first hitting time of the inside surface of $S_c \cup S_b$, $b>c$, for the ground-state process X; then

$$E[T(0^+,b)] = \frac{2ma^2}{\hbar} \int_0^{b/a} \{\frac{\sin hu}{u} - (1+u)e^{-u}\}^2 du. \tag{4.1}$$

Proof: Let $T_x[c,b] = \inf\{t\geq 0 : |X_t| \in (c,b)^c, X_0 = x\}$, and put $v(x)=E[T_x[c,b]]$. Then, using Dynkin's identity, as explained in Professor Pinsky's contribution to these proceedings, v is the unique solution of the equation

$$\frac{\sigma^2}{2}v''(x) + a(x)v'(x) = -1, \tag{4.2}$$

with $v(c)= v(b) = 0$, where $\sigma^2=\frac{\hbar}{m}$ and $a(x)$ is given by

$$a(x) = \frac{\hbar}{m}\left(\frac{1}{|X_t|} - \frac{1}{a}\right). \tag{4.3}$$

This gives

$$v(x) = E[T_x[c,b]] = -\int_x^b [\frac{d}{x^2}\exp(\frac{2x}{a}) + \frac{2m}{\hbar}(\frac{a}{2}+\frac{a^2}{2x}+\frac{a^3}{4x^2})]dx, \tag{4.4}$$

with $d=d(c)$ determined by the condition $v(c)=0$. Hence

$$d(c) = -\frac{m}{\hbar}\int_c^b (a + \frac{a^2}{x} + \frac{a^3}{2x^2})dx \Big/ \int_c^b \frac{1}{x^2}\exp(\frac{2x}{a})dx.$$

It is not difficult to show that $E[T(b)] = E[T(0^+,b)]$, so we need to calculate $c(0^+)$. Writing

$$d(c) = \{-\frac{m}{\hbar}\int_c^b a\,dx - \frac{ma^3}{2\hbar}\int_c^b \frac{2}{ax}+\frac{1}{x^2}\,dx\} \Big/$$

$$\left\{\int_c^b(\frac{1}{x^2}\exp(\frac{2x}{a}) - \frac{2}{ax} - \frac{1}{x^2})dx + \int_c^b(\frac{2}{ax} + \frac{1}{x^2})dx\right\} \tag{4.5}$$

we see that $d(0^+) = -\frac{ma^3}{2\hbar}$. This is precisely the value of d required to make $v'(0^+)$ finite. Repeated integration by parts completes the proof

It is interesting to compute the orders of magnitude of the quantiti involved here; as $\frac{b}{a} \sim \infty$. Crudely putting $e^{2u}/u^2 \sim e^{2u}$ as $u \sim \infty$, we have

$$\bar{T}(b) = E[T(0^+,b)] \sim \frac{ma^2}{4\hbar}\exp\left(\frac{2b}{a}\right).$$

If $\bar{T}(b)$ is of the order of one year, then b is of the order of 28 Bohr radii; if $\bar{T}(b)$ is of the order of the age of the universe (10^{10} years), then b is of the order of 40 Bohr radii. Hence, on the basis of stochastic mechanics, the expected first time the electric dipole moment of the Hydrogen atom, in its ground state, exceeds 28 ae is of the order of one year. These results underline the very high accuracy required to check the predictions experimentally.

Next we determine the distribution of the first hitting time by computing its Laplace transform. We require some notation: define $\bar{v}_i(x,E)$, i=1,2, by

$$\bar{v}_1(x,E) = \frac{1}{x}e^{x}W_{(1+2E)^{-\frac{1}{2}},\frac{1}{2}}(2(1+2E)^{\frac{1}{2}}x), \tag{4.6}$$

and

$$\bar{v}_2(x,E) = \frac{1}{x}e^{x}M_{(1+2E)^{-\frac{1}{2}},\frac{1}{2}}(2(1+2E)^{\frac{1}{2}}x), \tag{4.7}$$

where $W_{\ell,\frac{m}{2}}$ and $M_{\ell,\frac{m}{2}}$ are the confluent hypergeometric solutions of Whittaker's differential equation:

$$v''_{\ell,\frac{m}{2}} + \left\{-\frac{1}{4} + \frac{\ell}{x} + \frac{(1-m^2)}{4x^2}\right\}v_{\ell,\frac{m}{2}} = 0, \tag{4.8}$$

and $\bar{v}_1$ is singular at x=0 and $\bar{v}_2$ is regular at x=0

To avoid notational complications we pass to dimensionless variables, eliminating Planck's constant $\hbar$ and the reduced mass m from the equations:

Proposition 5

Let $T_x(b)$ be the first hitting time of the inside surface of the sphere S_b for the diffusion associated with the ground-state of the Hydrogen atom starting at x:

$$T_x(b) = \inf\{t \geq 0 : |X_t| \in [b,\infty]\}. \tag{4.9}$$

Let $\bar{v}(x,E)$ be defined by

$$\bar{v}(x,E) = \int_0^\infty e^{-Et}P[T_x(b) > t]\,dt.$$

Then for x in (0,b) we have

$$\bar{v}(x,E) = -\frac{\Gamma(1-(1+2E)^{-\frac{1}{2}})}{(1+2E)^{\frac{1}{2}}}\{\int_x^b w_1(y)y^2e^{-2y}dy\, w_2(x) - w_1(x)\int_0^x w_2(y)y^2e^{-2y}dy\}\,, \tag{4.10}$$

where

$$\begin{bmatrix} w_1 \\ \\ w_2 \end{bmatrix} = \begin{bmatrix} \left(\frac{\bar{v}_2(b)}{\bar{v}_1(b)}\right)^{\frac{1}{2}} & -\left(\frac{\bar{v}_1(b)}{\bar{v}_2(b)}\right)^{\frac{1}{2}} \\ \\ 0 & -\left(\frac{\bar{v}_1(b)}{\bar{v}_2(b)}\right)^{\frac{1}{2}} \end{bmatrix} \begin{bmatrix} \bar{v}_1 \\ \\ \bar{v}_2 \end{bmatrix} \tag{4.11}$$

Proof: Let d be an open arcwise connected set in $\mathbb{R}^d$ and put

$T_x(D) = \inf\{ s \geq 0 : Z_x(s) \in D^C \}$, Z_x being a diffusion, starting at x, with generator L. Then it can be shown that

$$P[T_x(D)>t] = \exp(t L_D^*)1_D(x), \tag{4.12}$$

where 1_D is the characteristic function of D, and L_D^* is the adjoint of L_D, the Dirichlet form of the operator associated with the generator L with Dirichlet boundary conditions on the boundary of D. The desired result follows by taking Laplace transforms and carrying out some rather unpleasant calculations.

Remarks:

1. Since $L = -|f_E|(H-E)|f_E|^{-1}$, the last displayed equation gives

$$P[\bar{T}(D) > t] = \int_{\mathbb{R}} e^{-t(\lambda-E)} d_\lambda \| E(H_D)(\lambda)|f_E| \|^2,$$

where $E(H_D)(\lambda)$ is the spectral measure corresponding to the self-adjoint operator H_D with Dirichlet boundary conditions on the boundary of D. Using this fact, we can avoid the use of Laplace transforms.

2. The above analysis can be generalized to the excited states $f_{E'}$. We must restrict ourselves to working in a connected region S, bounded by a nodal surface N, in which $f_{E'}$ is nowhere zero. We impose Dirichlet boundary conditions on N. In the region S we study the diffusion X with drift (grad) (Re + Im) log $|f_{E'}1_S|$, having generator $L = |f_{E'}1_S|(H_D^N - E')|f_{E'}1_S|^{-1}$, where H_D^N is the self-adjoint Hamiltonian with Dirichlet boundary conditions on N. (See [8].) Note that $|f_{E'}1_S|$

is the ground-state wave-function of the Hamiltonian H_D^N with eigenvalue E'. The invariant measure for the process X is $|f_{E'1S}|^2(x)dx$. Putting

$$H_D = \lim_{\lambda\uparrow\infty} (H_D^N + \lambda 1_{S^c})$$

so that $H_D \geq E'$ in the sense of quadratic forms; in this situation we obtain

$$P[\overline{T}(D) \in (t,t+dt)] = (|f_{E'}|, (H_D-E')\exp(-t(H_D-E'))|f_{E'}|)dt,$$

where $(\cdot,\cdot)$ is the inner product in $L_2(D)$, and $\overline{T}(D)$ is the expected first hitting time of the boundary of $D \subset S$. (See [1].)

References

[1] A. Truman and J.T. Lewis: Stochastic Mechanics of the Bound States of the Hydrogen Atom, in course of preparation.

[2] E.B. Davies: Quantum Stochastic Processes, Commun. Math. Phys. 15, 277-304 (1969).

[3] N. Ikeda and S. Watanabe: Stochastic Differential Equations and Diffusion Processes. Amsterdam-Oxford-New York: North Holland 1981.

[4] A. Friedman: Partial Differential Equations of Parabolic Type. Englewood Cliffs, New Jersey: Prentice Hall 1964.

[5] E. Nelson, Dynamical Theories of Brownian Motion, Princeton: Princeton University Press 1967.

[6] B. Simon: Semiclassical Analysis of Low-Lying Eigenvalues, II, Cal. Tech. Preprint 1984.

[7] I. Davies and A. Truman: Laplace Asymptotic Expansions of Conditional Wiener Integrals and Generalized Mehler Kernel Formulae for Hamiltonians on $L^2(\mathbb{R}^n)$, J. Phys. A17, 2773-2789 (1984).

[8] S. Albeverio and R. Høegh-Krohn: A Remark on the Connection Between Stochastic Mechanics and the Heat Equation, J. Math. Phys. 15, 1745-1747 (1974).

[9] H.P. McKean: Stochastic Integrals. Academic Press, New York (1969).

NONSTANDARD ANALYSIS AND PERTURBATIONS OF THE LAPLACIAN ALONG BROWNIAN PATHS

by

Tom Lindstrøm
Department of Mathematics
University of Trondheim
N-7034 Trondheim-NTH, Norway

Introduction

During the last ten years or so, there has been an increasing interest in nonstandard methods in probability theory and mathematical physics, and a growing awareness that the theory of infinitesimals provides new and promising ways of modeling natural and social phenomena. Since most mathematicans are unfamiliar with even the fundamentals of nonstandard analysis, this development has created a problem of communication as people suddenly find themselves unable to read papers purporting to be in their own field. The problem should not be exaggerated; contrary to popular belief, nonstandard analysis is not a difficult theory based on deep and mysterious results of mathematical logic, but can be viewed as the result of a rather straightforward and natural limit construction. This is the approach I shall take in the present paper, and although reading these twenty pages wont't turn you into an instant expert, I hope to give you a feeling for the basic ideas of the subject and the ways in which they can be applied. For additional information, you should take a look at the books [5], [8], [10], [12], [15], [14], [2] and also the excellent survey papers [4] and [7].

I have divided this paper into two parts. In the first, I'll not only show you a simple and natural construction of the nonstandard real numbers (infinitesimals and all), but also how they are used to model Brownian motion; in the second part, I'll give an application to the perturbation theory of Schrödinger operators. If this application is what mainly interests you, there is a self-contained description of the problem and the result at the beginning of part II.

I. Nonstandard analysis

A. Construction of $^*\mathbf{R}$

I have promised to construct a set $^*\mathbf{R}$ of nonstandard real numbers in a "simple and natural" way. If you are a little cynical, you might say that what is "simple and natural" in mathematics is what looks pretty much like something you have seen before, and that my task therefore is to make the construction of $^*\mathbf{R}$ as similar as possible to something you are already familiar with. Well, what about the construction of the reals from the rationals using Cauchy-sequences?

Recall how this is done; let C be the set of all rational Cauchy-sequences, and define an equivalence relation $\equiv$ on C by

(1) $\{a_n\} \equiv \{b_n\}$ if and only if $\lim_{n\to\infty}(a_n - b_n) = 0$.

The set of all real numbers is just the set $\mathbf{R} = C/\equiv$ of all equivalence classes. To define algebraic operations on $\mathbf{R}$, let $\langle a_n \rangle$ denote the equivalence class of the sequence $\{a_n\}$, and define addition and multiplication componentwise

(2) $\langle a_n\rangle + \langle b_n\rangle = \langle a_n+b_n\rangle$; $\langle a_n\rangle\cdot\langle b_n\rangle = \langle a_n\cdot b_n\rangle$.

To order $\mathbf{R}$, simply let $\langle a_n\rangle < \langle b_n\rangle$ if $a_n < b_n$ for all sufficiently large n. Finally, we can identity the rationals with a subset of $\mathbf{R}$ through the embedding

(3) $a \to \langle a,a,a,a,\ldots.\rangle$.

The construction of a set $^*\mathbf{R}$ of nonstandard reals follows exactly the same lines. Starting with the set S of all sequences of real numbers, I shall introduce an equivalence relation $\sim$ on S, and define $^*\mathbf{R}$ as the set $S/\sim$ of all equivalence classes. If as before $\langle a_n\rangle$ denotes the equivalence class of the sequence $\{a_n\}$, the algebraic operations are defined componentwise as in (2) , and I will also introduce an ordering on $^*\mathbf{R}$ which turns it into an ordered field. Finally, $\mathbf{R}$ will be identified with a subset of $^*\mathbf{R}$ through the embedding $a \to \langle a,a,a,\ldots\rangle$.

Before defining the equivalence relation $\sim$ above, it may be wise to say a few words about the philosophy behind the construction. In creating the reals from the rationals, one is interested in constructing limit points for all "naturally" convergent sequences. Since the limit is all one cares about, it is convenient to identify as <u>many</u> sequences as possible; i.e. all those which converge to the same "point". No attention is paid to the rate of convergence; hence the two sequences $\{\frac{1}{n}\}$ and $\left\{\frac{1}{n^2}\right\}$ are both identified with the same number 0 although they converge at quite different rates. In creating $^*\mathbf{R}$ from $\mathbf{R}$, one is interested in constructing a rich and well-organized algebraic structure which encodes not only the <u>limit</u> of a sequence but also its <u>rate of convergence</u>. To achieve this one must reverse the strategy above and identify as <u>few</u> sequences as possible.

This sounds rather silly; "identifying as few sequences as possible", doesn't that just mean the trivial identification $\{a_n\} \sim \{b_n\}$ if $\{a_n\} = \{b_n\}$? Well, it doesn't if you want to keep a decent algebraic structure on your model:

<u>I-1Example</u>: Let $\{a_n\} = \{1,0,1,0,1,\dots\}$ and $\{b_n\} = \{0,1,0,1,0,\dots\}$, then $\{a_n\}\cdot\{b_n\} = \{0,0,0,\dots\} = 0$, although $\{a_n\}$ and $\{b_n\}$ are both nonzero. Thus if you don't identify any sequences, you get a structure with zero divisors.

The idea is to make the equivalence relation $\sim$ just strong enough to avoid the problem of zero divisors. Before I can give the definition, I have to fix a finitely additive measure on $\mathbf{N}$ with certain special properties:

<u>I-2 Definition</u>: Throughout this paper m will denote a (fixed) finitely additive measure on the set $\mathbf{N}$ of positive integers such that:

(i) For all $A \subset \mathbf{N}$, $m(A)$ is defined and is either 0 or 1.

(ii) $m(\mathbf{N}) = 1$, and $m(A) = 0$ for all finite A.

Note that m divides the subsets of $\mathbf{N}$ into two classes; the "big" ones with measure one, and the "small" ones with measure zero; and that all finite sets are "small". That finitely additive measures of this kind exist, is easily proved by Zorn's lemma.

<u>I-3 Definition</u>: Let $\sim$ be the equivalence relation on the set S of all sequences of real numbers defined by

$$\{a_n\} \sim \{b_n\} \text{ if and only if } m\{n \mid a_n = b_n\} = 1 ,$$

i.e. if $\{a_n\}$ equals $\{b_n\}$ almost everywhere.

Having defined the equivalence relation $\sim$, I can now do as promised and let the set $^{*}\mathbf{R} = S/\sim$ be my set of nonstandard reals. If $\langle a_n \rangle$ denotes the equivalence class of the sequence $\{a_n\}$, define addition and multiplication by

$$\langle a_n \rangle + \langle b_n \rangle = \langle a_n + b_n \rangle \ ; \ \langle a_n \rangle \cdot \langle b_n \rangle = \langle a_n \cdot b_n \rangle \ ,$$

an introduce an ordering by

$$\langle a_n \rangle < \langle b_n \rangle \ \text{ iff } \ m\{n \mid a_n < b_n\} = 1.$$

I shall leave the routine exercise of verifying that $^{*}\mathbf{R}$ is an ordered field with zero element $0 = \langle 0,0,0,\ldots \rangle$ and unit $1 = \langle 1,1,1,\ldots \rangle$ to you. But let me at least show why the problem of zero divisors has disappeared as this is quite instructive:

Assume that $\langle a_n \rangle \langle b_n \rangle = 0$, i.e. $m\{n \mid a_n \cdot b_n = 0\} = 1$. Since $\{n \mid a_n \cdot b_n = 0\} = \{n \mid a_n = 0\} \cup \{n \mid b_n = 0\}$, either $\{n \mid a_n = 0\}$ or $\{n \mid b_n = 0\}$ has measure one, and thus either $\langle a_n \rangle = 0$ or $\langle b_n \rangle = 0$. Note that the conditions on m are exactly right for this argument to work.

As already indicated, the map $a \to \langle a,a,a,\ldots \rangle$ is an injective, orderpreserving homomorphism identifying $\mathbf{R}$ with a subset of $^{*}\mathbf{R}$, and I shall not distinguish between a and $\langle a,a,a,\ldots \rangle$. But what do the other elements in $^{*}\mathbf{R}$ look like? After all, nonstandard analysis is supposed to provide a consistent way of introducing infinitesimals in analysis, and this far I haven't shown you a single one! Let us first fix the terminology:

I-4 Definition: (a) An element $x \in {}^{*}\mathbf{R}$ is <u>infinitesimal</u> if $-a < x < a$ for all positive, real numbers a.

(b) An element $x \in {}^{*}\mathbf{R}$ is <u>finite</u> if $-a < x < a$ for some positive, real number a.

Three examples of infinitesimals are 0, $\delta_1 = \langle \frac{1}{n} \rangle$ and $\delta_2 = \langle \frac{1}{n^2} \rangle$. To check that, say, δ_1 is infinitesimal, note that for any positive $a \in \mathbf{R}_+$, the set $\{n \mid -a < \frac{1}{n} < a\}$ contains all but a finite number of n's, and hence has measure one. Observe also that since $\delta_1 \neq \delta$, the two sequences $\{\frac{1}{n}\}$ and $\{\frac{1}{n^2}\}$ converging to zero at different rates are represented by different infinitesimals. Examples of infinite numbers are $\delta_1^{-1} = \langle n \rangle$ and $\delta_2^{-1} = \langle n^2 \rangle$.

It is easy to check that the arithmetic rules one would expect really hold; e.g. is the sum of two infinitesimals an infinitesimal and the product of a noninfinitesimal number and an infinite one is infinite.

More interesting is the following observation which shows that the finite part of $^*\mathbf{R}$ has a very simple structure.

<u>I-5 Lemma</u>. Any finite $x \in {}^*\mathbf{R}$ can be written uniquely as a sum $x = a + \varepsilon$, where $a \in \mathbf{R}$ and ε is infinitesimal.

Proof: The uniqueness is obvious; if $x = a_1 + \varepsilon_1 = a_2 + \varepsilon_2$, then $a_1 - a_2 = \varepsilon_2 - \varepsilon_1$, and since this quantity is both real and infinitesimal, it must be zero.

For the existence, let $a = \sup\{b \in \mathbf{R} \mid b < x\}$; since x is finite, a exists. I must show that $x - a$ is infinitesimal. Assume not, then there is a real number r such that $0 < r < |x - a|$. If $x - a$ is positive, this implies that $a + r < x$, contradicting the choice of a. If $x - a$ is negativ, we get $x < a - r$, also contradicting the choice of a.

<u>I-6 Definition</u>. For each finite $x \in {}^*\mathbf{R}$, the unique real number a such that $x - a$ is infinitesimal is called the <u>standard part</u> of x and denoted by $\mathrm{st}(x)$ or $^\circ x$.

Let us also agree that $x \approx y$ means that x and y are <u>infinitely close</u>; i.e. $x - y$ is infinitesimal.

B Internal sets and functions.

One of the first things you do when you have introduced a new mathematical structure is to look for the classes of "nice" subsets and functions (such as open sets and continuous functions in topology, measurable sets and integrable functions in measure theory). In non-standard analysis the "nice" sets and functions are called internal, and they arise in the following fashion:

A sequence $\{A_n\}$ of subsets of $\mathbf{R}$ defines a subset $\langle A_n \rangle$ of $^*\mathbf{R}$ by

$$\langle x_n \rangle \in \langle A_n \rangle \text{ if and only if } m\{n \mid x_n \in A_n\} = 1.$$

Any subset of $^*\mathbf{R}$ which can be obtained in this way is called <u>internal.</u> Similarly, a sequence $\{f_n\}$ of functions from $\mathbf{R}$ to $\mathbf{R}$ defines a function $\langle f_n \rangle : {}^*\mathbf{R} \to {}^*\mathbf{R}$ through

$$\langle f_n \rangle (\langle x_n \rangle) = \langle f_n(x_n) \rangle,$$

and functions of this kind are called <u>internal</u>.

<u>I-7 Example</u>: (a) If $a = \langle a_n \rangle$ and $b = \langle b_n \rangle$ are two elemetns of ${}^*\mathbf{R}$, then the interval $[a,b] = \{x \in {}^*\mathbf{R} \mid a \leq x \leq b\}$ is internal since it equals $\langle [a_n, b_n] \rangle$.

(b) If $c = \langle c_n \rangle$ is in ${}^*\mathbf{R}$, the function $\sin(cx)$ is an internal function defined by $\sin(cx) = \langle \sin(c_n x_n) \rangle$.

(c) There are many sets and functions which are not internal, e.g. the set of all infinitesimals and its indicator function.

What is important about the internal sets and functions is that their product-like structure makes it possible to lift operations componentwise from $\mathbf{R}$ to ${}^*\mathbf{R}$; as an example, define the nonstandard integral $\int_A f\,dx$ where $A = \langle A_n \rangle$ is an internal set and $f = \langle f_n \rangle$ is an internal function by

$$(4) \quad \int_A f\,dx = \langle \int_{A_i} f_i\,dx \rangle .$$

This new integral inherits most of the properties of the standard integral; both general statements such as $\int_A (f+g)\,dx = \int_A f\,dx + \int_A g\,dx$ and more specific ones such as

$$\int_a^b \sin(cx)\,dx = \frac{1}{c}\cos(ca) - \frac{1}{c}\cos(cb)$$

for $a,b,c \in {}^*\mathbf{R}$, remain true.

Another principle which carries over from $\mathbf{R}$ to the internal sets is the least upper bound formulation of the completeness axiom; if $A = \langle A_i \rangle$ is an internal set bounded by some element $a = \langle a_i \rangle$ in ${}^*\mathbf{R}$, then almost all the A_i's are bounded by the corresponding a'_is, and it is easy to check that $b = \langle \sup(A_i) \rangle$ is a least upper bound for $\langle A_i \rangle$. I should point out at this stage that the least upper bound principle does not hold for <u>all</u> subsets of ${}^*\mathbf{R}$; if it did, ${}^*\mathbf{R}$ would satisfy all the axioms for the real numbers and hence be isomorphic to them. That ${}^*\mathbf{R}$ is not complete often worries people in the beginning, but they soon realize that the completeness of $\mathbf{R}$ and the least upper bound principle for internal sets suffice.

It is easy to see that the family of internal sets is closed under finite Boolean operations and thus form an algebra; indeed

$$(5) \quad \langle A_n \rangle \cap \langle B_n \rangle = \langle A_n \cap B_n \rangle; \quad \langle A_n \rangle \cup \langle B_n \rangle = \langle A_n \cup B_n \rangle; \quad C\langle A_n \rangle = \langle CA_n \rangle .$$

The next result states another basic fact about internal sets. Note the similarity with compacts.

I-8 Theorem (Keisler): Let $\{A^i\}_{i\in\mathbf{N}}$ be a sequence of internal sets such that $\bigcap_{i\leq I} A^i \neq \emptyset$ for all $I\in\mathbf{N}$. Then $\bigcap A^i \neq \emptyset$.

Proof: Each A^i is of the form $\langle A_n^i\rangle$, and since $A^1 \neq \emptyset$, I can clearly assume that $A_n^1 \neq \emptyset$ for all n. By (5), $\langle \bigcap_{i\leq I} A_n^i\rangle = \bigcap_{i\leq I} A^i \neq \emptyset$, and thus

(6) $m\{n \mid \bigcap_{i\leq I} A_n^i \neq \emptyset\} = 1$

for all $I\in\mathbf{N}$. For each n, let

$$I_n = \max\{I \mid \bigcap_{i\leq I} A_n^i \neq \emptyset \text{ and } I\leq n\},$$

and pick an element $x_n \in \bigcap_{i\leq I_n} A_n^i$. It suffices to prove that $\langle x_n\rangle \in A^I$ for all I, and this follows from (6) since

$$\{n \mid x_n \in A_n^I\} \supset \{n \mid I_n \geq I\} =$$

$$\{n \mid n\geq I\} \cap \{n \mid \bigcap_{i\leq I} A_n^i \neq \emptyset\},$$

where $\{n \mid n\geq I\}$ has finite complement and thus measure one.

A quite curious (and, it will turn out, extremely useful) consequence of this result is that the algebra of internal sets is as far from being a σ-algebra as an algebra can possibly be.

I-9 Corollary: If $\{A_n\}_{n\in\mathbf{N}}$ is a sequence of internal sets, then the union $\bigcup_{n\in\mathbf{N}} A_n$ is internal if and only if it equals $\bigcup_{n\leq N} A_n$ for some $N\in\mathbf{N}$.

Proof: Assume that $A = \bigcup_{n\in\mathbf{N}} A_n$ is internal. Then all the sets $A - A_n$ are internal, and clearly $\bigcap_{n\in\mathbf{N}} (A\setminus A_n) = \emptyset$. By the theorem there must be an N such that $\bigcap_{n\leq N} (A\setminus A_n) = \emptyset$, and consequently $A = \bigcup_{n\leq N} A_n$.

Internal sets and functions which deserve special mention are the standard ones. The internal set $\langle A_n\rangle$ is called standard if (almost) all the A_n's are equal to the same set A -in which case $\langle A_n\rangle$ is denoted by $^*A$. Similarly, a standard function $^*f$ is an internal function $\langle f_n\rangle$ where (almost) all the f_n's are equal to f. In this way each subset $A\subset\mathbf{R}$ has a nonstandard version $^*A\subset {}^*\mathbf{R}$. Note that $^*A$ is usually much richer than A; e.g. will $^*(a,b)$ contain not only all real numbers between a and b, but even all nonstandard numbers with the same property. Similarly is the nonstandard version $^*f$ of a function $^*f$ an extension of f to $^*\mathbf{R}$, thus

${}^*f(x) = f(x)$ for all $x \in \mathbf{R}$.

An important standard set is ${}^*\mathbf{N}$, the set of all nonstandard natural numbers. In addition to the ordinary natural numbers, this set contains infinite elements such as $\langle 1,2,3,4,\ldots\rangle$

C. Hyperfinite sets and Brownian motion.

One of the most interesting discoveries of nonstandard analysis is the existence of hyperfinite sets. These are infinite sets with most of the properties of finite sets, and they are convenient tools for modeling a wide variety of phenomena. As a simple first example, I'll show you how to use them to construct Brownian motion.

A hyperfinite set is simply an internal set $A = \langle A_n\rangle$ where (almost all) the A_n's are finite. The internal cardinality $|A|$ of A is the nonstandard natural number $\langle |A_n|\rangle$, where $|A_n|$ is the number of elements in A_n.

I-10 Example: If $N = \langle N_n\rangle$ is a nonstandard natural number, then the set $T = \{0, \frac{1}{N}, \frac{2}{N}, \ldots, \frac{N-1}{N}, 1\}$ is hyperfinite with internal cardinality $N+1$. This is because $T = \langle T_n\rangle$, where $T_n = \{0, \frac{1}{N_n}, \frac{2}{N_n}, \ldots, 1\}$.

As an illustration of how finite notions can be extended to hyperfinite sets, take an internal function $f = \langle f_n\rangle$ and a hyperfinite set $A = \langle A_n\rangle$, and define the sum of f over A by

$$(7) \qquad \sum_{a \in A} f(a) = \langle \sum_{a_n \in A_n} f_n(a_n)\rangle .$$

If T is as in the example above and $g : \mathbf{R} \to \mathbf{R}$ is a function, then according to this definition

$$\sum_{t \in T} {}^*g(t)\frac{1}{N} = \langle \sum_{t_n \in T_n} g(t_n)\frac{1}{N_n}\rangle .$$

If g is continuous, the sequence on the right converges to $\int_0^1 g(t)dt$, and thus

$$\int_0^1 g(t)dt = st\left(\sum_{t \in T} {}^*g(t)\frac{1}{N}\right);$$

the Riemann integral is nothing but a hyperfinite sum.

Let us take a look at Brownian motion. Fix an infinitely large integer $N = \langle N_n\rangle$ and let $T = \langle T_n\rangle$ be as in Example 1 - 10. The idea is to obtain Brownian motion as the standard part of a random walk with infinitesimal steps. To be more precise, let Ω be the set of all internal maps $\omega : T \to \{-1,1\}$, and define a map $\chi : \Omega \times T \to {}^*\mathbf{R}$ by

$$(8) \quad \chi(\omega,t) = \sum_{s=0}^{t-1/N} \frac{\omega(s)}{\sqrt{N}} .$$

Since I'm summing an internal function $\frac{\omega(s)}{\sqrt{N}}$ over a hyperfinite set $\{0,\frac{1}{N},\frac{2}{N},\ldots,t-\frac{1}{N}\}$, this definition makes sense. To turn χ into a realvalued process $b : \Omega \times [0,1] \to \mathbf{R}$, put

$$(9) \quad b(\omega,t) = st(\chi(\omega,\tilde{t}))$$

where $\tilde{t}$ denotes the element in T to the immediate left of t.

To claim that b is a Brownian motion, I need a measure on Ω. This set consists of all internal functions $\omega : T \to \{-1,1\}$ where T has internal cardinality $N+1$, and hence it ought to be a hyperfinite set of cardinality 2^{N+1}. And so it is; in fact, it is trivial to check that $\Omega = \langle \Omega_n \rangle$, where Ω_n is the set of all maps $\omega_n : T_n \to \{-1,1\}$. Since Ω is hyperfinite, I can for all internal $A \subset \Omega$ define

$$P(A) = \frac{|A|}{2^{N+1}} .$$

This is a ${}^*\mathbf{R}$-valued, finitely additive measure on the algebra of internal subsets of Ω. I can turn it into a $\mathbf{R}$-valued, finitely additive measure ${}^\circ P$ simply by taking standard parts: ${}^\circ P(A) = st(P(A))$. By Corollary 1 - 9 the conditions of Caratheodory's extension theorem are trivially satisfied, and hence ${}^\circ P$ can be uniquely extended to a measure on the σ-algebra generated by the internal sets. The completion of this measure I'll denote by $L(P)$ and refer to as the <u>Loeb-measure</u> of P.

With the probability measure $L(P)$ on Ω, the process b is a Brownian motion. Although I shan't prove this in detail, I would like to sketch a quick proof of the a.s. path continuity as this is the only really nontrivial part. In the proof, E will denote expectation with respect to the internal measure P; i.e

$$E(F) = 2^{-(N+1)} \sum_{\omega \in \Omega} F(\omega) .$$

Since Ω is hyperfinite, this makes sense for all internal functions F.

The argument is based on the well-known expression

$$E((\chi(t)-\chi(s))^4) = 3(t-s)^2 - \frac{2|t-s|}{N} < 3(t-s)^2$$

for the fourth moment of the random walk χ. This formula holds for all finite random walks, and by now you should be able to figure out why it also holds for the hyperfinite walk χ.

Here comes the proof: For each pair $(m,n) \in \mathbf{N}^2$, define a "bad" set $\Omega_{m,n}$ by

$$\Omega_{m,n} = \{\omega \mid \exists i < m \exists s \in T \cap [\tfrac{i}{m},\tfrac{i+1}{m}](|\chi(\omega,\tfrac{i}{m}) - \chi(\omega,s)|^4 \geq \tfrac{1}{n})\}.$$

To show that b is continuous, it suffices to prove that for each n, ${}^\circ P(\Omega_{m,n}) \to 0$ as $m \to \infty$. Now

$$(10) \quad P(\Omega_{m,n}) \leq \sum_{i=0}^{m-1} P\{\omega \mid \exists s \in T \cap [\tfrac{i}{m},\tfrac{i+1}{m}](|\chi(\omega,\tfrac{i}{m}) - \chi(\omega,s)|^4 \geq \tfrac{1}{n})\}$$
$$\leq 2\sum_{i=0}^{m-1} P\{\omega \mid |\chi(\omega,\tfrac{i}{m}) - \chi(\omega,\tfrac{i+1}{m})|^4 \geq \tfrac{1}{n}\},$$

where the last inequality comes from the following well-known reflection argument: If $|\chi(\omega,\frac{i}{m}) - \chi(\omega,\frac{i+1}{m})|^4 < \frac{1}{n}$, but there is an $s \in [\frac{i}{m},\frac{i+1}{m}]$ such that $|\chi(\omega,\frac{i}{m}) - \chi(\omega,s)|^4 \geq \frac{1}{n}$, let s_ω be the smallest such s. Define $\tilde{\omega}$ to be the reflected path

$$\omega(r) = \begin{cases} \omega(r) & \text{if } r < s_\omega \\ -\omega(r) & \text{if } r \geq s_\omega \end{cases},$$

then $|\chi(\omega,\frac{i}{m}) - \chi(\omega,\frac{i+1}{m})|^4 \geq \frac{1}{n}$. Since there is an internal one-to-one correspondence between reflected and unreflected paths, (10) follows. The rest is just an easy computation.

$$P(\Omega_{m,n}) \leq 2\sum_{i=0}^{m-1} P\{\omega \mid |\chi(\omega,\tfrac{i}{m}) - \chi(\omega,\tfrac{i+1}{m})|^4 \geq \tfrac{1}{n}\}$$
$$\leq 2\sum_{i=0}^{m-1} nE(|\chi(\omega,\tfrac{i}{m}) - \chi(\omega,\tfrac{i+1}{m})|^4)$$
$$\leq 6n\sum_{i=0}^{m-1} \frac{1}{m^2} = \frac{6n}{m} \to 0 \quad \text{as} \quad m \to 0.$$

I have presented this proof in such detail because it illustrates one of the most interesting features of the nonstandard construction of Brownian motion; namely, the way in which combinatorial arguments (such as the reflection principle above) apply. In a sense, b has the Donsker invariance principle built into it; it is at the same time both a Brownian motion and a random walk. This possibility of doing combinatorics on infinite structures is one of the great assets of hyperfinite models in general.

D. A few remarks on nonstandard methods.

In order to "demystify" the subject, I have slightly misrepresented what nonstandard analysis is like. Just as nobody studies real analysis by considering equivalence classes of Cauchy sequences of rationals, nobody works in nonstandard analysis by constantly referring to equivalence classes of real sequences. If you take a closer look at what these sequences have been used for, you will realize that it has been to carry over operations and results from **R** to ***R**. It turns out that there is a theorem (aptly named "the transfer principle") which classifies once and for all which properties can be carried over in this way and with what consequences. Although it is not at all hard to prove, I have not presented this result here as it is phrased in terms of the socalled "first order languages" of mathematical logic which it takes a while to explain (see any of the books on nonstandard analysis listed in the references). Once you have the transfer principle, you can turn the situation around and use it to give an axiomatic description of what a nonstandard number system should be. Based on these axioms you can then develop the subject without any references to equivalence classes of sequences. I probably should point out that the axiom systems I'm talking about have other models than the structure ***R** constructed above (examples are appropriate quotient spaces $\mathbf{R}^A/\sim$ for sets A richer than **N**), and that there thus is no unique set of nonstandard real numbers. Although some people seem to find this lack of uniqueness unsatisfactory for aesthetical reasons (and others use it as a pretext to turn philosophers [6]), it is not of any great practical consequence.

As everybody knows, the construction of **R** from **Q** generalizes to the construction of a completion for any given metric space. Is there a similar generalization of the construction of ***R** from **R**? The answer is "yes" - and very much so. In fact, given any set M, you can obtain a nonstandard version *M by simply following the recipe in section A. The reason why the nonstandard extension method is more general than the completion procedure, is, of course, that while in the latter case the equivalence relation on the space of sequences is defined in terms of the metric (and thus presupposes a metric structure), it's in the former defined solely in terms of the measure m on **N** (and thus doesn't require any kind of structure on M). In the second part of this paper, I'll make frequent use of the nonstandard extension $^*\mathbf{R}^d$ of $\mathbf{R}^d$.

Nothing of what I have told you so far is new. A version of the socalled ultrapower construction (i.e. the "equivalence classes of sequences" method that I have been explaining) was used already by Th. Skolem [13] in the first construction of a set of nonstandard natural numbers, and although A. Robinson often seemed to prefer to obtain $^*\mathbf{R}$ by a reference to the compactness theorem of first order logic (see e.g. [12]), the systematic application of ultrapowers in nonstandard analysis was soon exploited by W.A.J. Luxemburg [11]. The hyperfinite model of Brownian motion and the underlying measure construction are more recent; they are due to R.M. Anderson [3] and P.A. Loeb [9].

II. Perturbations of the Laplacian along Brownian paths

A. The problem and the results.

The purpose of this second part of the paper is to give an impression of how nonstandard methods are applied. The results I am going to present are the outcome of joint work with S. Albeverio, J.E. Fenstad and R. Höegh-Krohn, and a more general and detailed account will appear in our book [2]. There is an alternative approach using Fourier analysis and ultraviolet cutoffs which has been developed by some of us in collaboration with W. Karwowski ; see [1] for a discussion.

I'll first give a description of the problem and the results in entirely standard terms. Assume that μ is a completed Borel measure on a Hausdorff space X, and that B is a closed subset of X. Let E be a symmetric, bilinear form on $L^2(X,\mu)$, and assume that E is closed and bounded from below.

<u>II-1 Definition</u>: A symmetric, bilinear form F on $L^2(X,\mu)$ is a <u>perturbation of</u> E <u>supported on</u> B if

(i) F is closed and bounded from below, and $D[F] \supseteq D[E]$.

(ii) There is an $f \in D[E]$ such that $F(f,f) \neq E(f,f)$.

(iii) If $f \in D[E]$ vanishes in a neighbourhood of B, then $F(f,g) = E(f,g)$ for alle $g \in D[E]$.

I'm interested in the existence of perturbations supported on B when B has measure zero. Although the methods apply to all closed sets B and fairly general Markov forms E on second countable, locally compact spaces X, I'll restrict myself in this note to the situation where E is the closure of

$$(11)\quad E(f,g) = \frac{1}{2}\int_{\mathbf{R}^d} \nabla f\,\nabla g\,dx$$

in $L^2(\mathbf{R}^d,dx)$, and B is a Brownian path in $\mathbf{R}^d$. If you think of the Brownian path as a primitive model for a polymer molecule, you may consider a perturbation of E supported on B as the Hamiltonian form of a quantum mechanical particle interacting with the molecule through a short range interaction. I'll say a few more words about the relationship to more sophisticated polymer models and quantum fields at the end of the paper.

There is an obvious strategy for constructing perturbations of E supported on B; just let ρ and $\tilde{\lambda}$ be a probability measure and a bounded function on B, respectively, and define

$$(12)\quad F(f,g) = \frac{1}{2}\int_{\mathbf{R}^d} \nabla f\,\nabla g\,dx + \int_B \tilde{\lambda} fg\,d\rho$$

for all $f,g \in C_0^1(\mathbf{R}^d)$. As $\tilde{\lambda}$ and ρ vary, one would expect this formula to produce a large class of perturbations of E supported on B. There is one problem, however; F need not be lower bounded and closable. On the other hand, even if all the F's are unclosable, there may still be perturbations of E supported on B; it's known, for instance, that if B is a finite set (and not a Brownian path), then E has a perturbation supported on B if and only if $d \leq 3$, but that a perturbation can only be of the form (12) if $d = 1$.

Since B is a d-dimensional Brownian path, i.e.

$$B = \{b(\omega,t) \in \mathbf{R}^d \mid t \in [0,1]\}$$

for some $\omega \in \Omega$, the natural measure ρ to use on B is the one inherited from Lebesgue measure on $[0,1]$:

$$\rho(A) = dx\{t \in [0,1] \mid b(\omega,t) \in A\}.$$

I can now state the results I want to discuss:

<u>II-2 Theorem</u>: For almost all ω, the following holds:

(a) If $d \leq 5$, there is a perturbation of E supported on B

(b) If $d \leq 3$, the form (12) is closable for all finite Borel functions λ.

When $d = 4,5$, the perturbations of E will, in a certain sense, be obtained from (12) by an infinitesimal choice of $\tilde{\lambda}$.

<u>B. Hyperfinite quadratic forms.</u>

To prove the theorem above by nonstandard methods, I first have to rephrase the problem in nonstandard terms. Beginning with the Brownian motion, let $\chi_1, \chi_2, \ldots, \chi_d$ be d independent copies of the hyperfinite random walk defined in Part I, and put $\chi = (\chi_1, \chi_2, \ldots, \chi_d)$. If $\frac{1}{N}$ is the size of the time increments, let $\delta = \frac{1}{\sqrt{N}}$ and note that χ is a random walk on the lattice

$$\Lambda = \{(n_1\delta, n_2\delta, \ldots, n_d\delta) \mid n_i \in {}^*\mathbf{Z}, |n_i| \leq \delta^{-1} \text{ for all } i\}.$$

I'll leave it to you to check that Λ is a hyperfinite set.

If Δ_δ is the discrete Laplacian on Λ, i.e.

$$\Delta_\delta f(i) = \delta^{-2}\Big(\sum_{|j-i|=\delta} f(j) - 2df(i)\Big)$$

where $|\cdot|$ is the maximum norm $|x| = \max_{i \leq d} |x_i|$, the form

(13) $$E(f,g) = -\frac{1}{2}\sum_{i\in\Lambda} \Delta_\delta f(i) g(i) \delta^d$$

- defined for all internal functions f and g - is a nonstandard counterpart of (11). To get a corresponding nonstandard representation of (12), let $\lambda : \Lambda \to {}^*\mathbf{R}$ be an internal function and define

(14) $$F(f,g) = -\frac{1}{2}\sum_{i\in\Lambda} \Delta_\delta f(i) g(i)\delta^d + \sum_{t\in T} \lambda(\chi(t)) f(\chi(t)) g(\chi(t))\Delta t,$$

where $\Delta t = \frac{1}{N}$.

The idea is to construct perturbations of E supported on B by turning the hyperfinite forms F into standard forms. Since the standard and nonstandard forms operate on different spaces, it is necessary first to establish a correspondece between the elements of these two spaces. For each $i = (n_1\delta, \ldots, n_d\delta) \in \Lambda$, let $[i]$ be the half-open cube

$$[i] = \{(x_1,x_2,\dots,x_d) \in \mathbf{R}^d \mid n_i\delta \leq x_i < (n_i+1)\delta \quad \text{for all } i\}$$

If $f \in L^2(\mathbf{R}^d,dx)$, an internal function $\tilde{f}:\Lambda \to {}^*\mathbf{R}$ is called a <u>lifting</u> of f if

$$\sum_{i\in\Lambda} \left| \delta^{-d} \int_{[i]} {}^*f(x)dx - \tilde{f}(i)\right|^2 \delta^d \approx 0.$$

Obviously all $f \in L^2(\mathbf{R}^d,dx)$ have liftings; just define $\tilde{f}$ by $\tilde{f}(i) = \delta^{-d} \int_{[i]} {}^*f(x)dx$.

To turn F into a standard form ${}^\circ F$ on $L^2(\mathbf{R}^d,dx)$, let

$${}^\circ F(f,f) = \inf\{\mathrm{st}(F(\tilde{f},\tilde{f})) \mid \tilde{f} \text{ is a lifting of } f\},$$

where the values $\pm\infty$ just mean that f is not in the domain of ${}^\circ F$. The following (quite easy) result from chapter 5 of [2] shows that this definition is the right one.

<u>II-3 Proposition</u>. Assume that there is a $z_0 \in \mathbf{R}$ such that $F(f,f) \geq z_0\|f\|^2$ for all internal f, where $\|f\| = \left(\sum_{i\in\Lambda} f(i)^2\delta^d\right)^{\frac{1}{2}}$. Then F is a closed, symmetric, lower bounded form.

Not very surprisingly, it turns out that ${}^\circ E = E$. Since I want to use F to construct a perturbation of E, I need a criterion which tells me when ${}^\circ F$ and E are different. It will be convenient to have this criterion expressed in terms of the resolvents of E and F, and let thus

$$G_z = (-\Delta_\delta - z)^{-1}$$

be the resolvent of E, and $\tilde{G}_z$ the resolvent of F. Another easy result from chapter 5 of [2] now says.

<u>II-4 Proposition</u>. Assume that $z_0 = \inf\{\mathrm{st}\, F(f,f) \mid f = 1\}$ is finite. Assume further that there exist a $z \in \mathbf{R}$, $z < z_0$, and an internal function f with the following properties: $E(f,f)$ and $F(f,f)$ are both finite, and $(G_z - \tilde{G}_z)f$ is a lifting of a nonzero function in $L^2(\mathbf{R}^d,dx)$. Then ${}^\circ E \neq {}^\circ F$.

The stage is now set for the proof of Theorem II-2. Throughout this argument ω will be a fixed, "typical" element of Ω, where "typical" means that it belongs to a set of measure one where all the claims I make hold.

To compute the resolvent $\tilde{G}_z$ of F, note that

$$F(f,g) = \sum_{i\in\Lambda}\left(-\Delta_\delta f(i) + \frac{\lambda(i)\nu(i)}{\delta^d} f(i)\right)g(i)\delta^d,$$

where $\nu(i) = |\{t \mid \chi(\omega,t) = i\}|\Delta t$. Thus

$$\tilde{G}_z = \left(-\Delta_\delta + \frac{\lambda\nu}{\delta^d} - z\right)^{-1} = G_z\left(1 + \frac{\lambda\nu}{\delta^d} G_z\right)^{-1} = G_z + G_z \sum_{l=1}^{\infty}\left(-\frac{\lambda\nu}{\delta^d} G_z\right)^l$$

If $G_z(x,y)$ denotes the kernel (or, if you like, matrix) of G_z, i.e.

$$G_z f(x) = \sum_{y\in\Lambda} G_z(x,y) f(y)\delta^d,$$

then for all internal f

(15) $$G_z\left(-\frac{\lambda\nu}{\delta^d} G_z\right)^l f(x) =$$

$$= (-1)^l \sum_{(y,x_1,\ldots,x_l)\in\Lambda^{l+1}} G_z(x,x_l)\frac{\lambda(x_l)\nu(x_l)}{\delta^d} G_z(x_l,x_{l-1})\ldots$$

$$\ldots \frac{\lambda(x_1)\nu(x_1)}{\delta^d} G_z(x_1,y)f(y)\delta^{d(l+1)} =$$

$$= (-1)^l \sum_{(y,x_1,\ldots,x_l)\in\Lambda^{l+1}} G_z(x,x_l)\lambda(x_l)G_z(x_l,x_{l-1})\ldots$$

$$\ldots \lambda(x_1)G_z(x_1,y)f(y)\delta^d\nu(x_1)\ldots\nu(x_l).$$

Let C denote the "Brownian path" $\{\chi(\omega,t) \mid t \in T\}$; the expression above can be written more compactly if we introduce the operators $\hat{G}_z : L^2(C,\nu) \to L^2(\Lambda,\delta^d)$, $\hat{G}_z^* : L^2(\Lambda,\delta^d) \to L^2(C,\nu)$, $G_z' : L^2(C,\nu) \to L^2(C,\nu)$ defined by

$$\hat{G}_z f(x) = \sum_{y\in C} G_z(x,y)f(y)\nu(y)$$

$$\hat{G}_z^* f(x) = \sum_{y\in\Lambda} G_z(x,y)f(y)\delta^d$$

$$G_z' f(x) = \sum_{y\in C} G_z(x,y)f(y)\nu(y).$$

In this notation (15) becomes

$$G_z\left(-\frac{\lambda\nu}{\delta^d}G_z\right)^l = (-1)^l\,\hat{G}_z(\lambda G_z')^{l-1}\lambda\hat{G}_z^*,$$

and thus

$$(16)\quad \tilde{G}_z = G_z - \hat{G}_z\left(\sum_{l=0}^{\infty}(-\lambda G_z')^l\right)\lambda\hat{G}_z^*$$
$$= G_z - \hat{G}_z(1+\lambda G_z')^{-1}\lambda\hat{G}_z^*$$
$$= G_z - \hat{G}_z(\tfrac{1}{\lambda}+G_z')^{-1}\hat{G}_z^*$$

I have said nothing about the convergence of the infinite series involved in this calculation, but it is easy to check that the final answer is correct provided $(\frac{1}{\lambda}+G_z')$ is a strictly negative operator (and $(\frac{1}{\lambda}+G_z')^{-1}$ thus exists). Since

$$(17)\quad \langle(\tfrac{1}{\lambda}+G_z')g,g\rangle_{L^2(C,\nu)} = \sum_{x\in C}\left[\frac{g(x)}{\lambda(x)}+\sum_{y\in C}G_z(x,y)g(y)\nu(y)\right]g(x)\nu(x) =$$
$$\sum_{x\in C}\left[\frac{1}{\lambda(x)}+\sum_{y\in C}G_z(x,y)\nu(y)\right]g(x)^2\nu(x) - \frac{1}{2}\sum_{x\in C}\sum_{y\in C}G_z(x,y)\left[g(x)-g(y)\right]^2\nu(x)\nu(y)$$

the operator $(\frac{1}{\lambda}+G_z')$ is strictly negative if

$$\frac{1}{\lambda(x)} < -\sum_{y\in C}G_z(x,y)\nu(y)$$

for all x.

I now choose a number $z_0 \in \mathbf{R}_-$ and assume that $\lambda(x)$ is chosen such that

$$(18)\quad h(x) = \frac{1}{\lambda(x)} + \sum_{y\in C}G_{z_0}(x,y)\nu(y)$$

is strictly negative. By what I have just shown, this guarantees that (16) holds for all $z \le z_0$. It is easy to check that it also implies that $F(f,f) \ge z_0\|f\|^2$ for all f. Note that although the quantity $\sum_{y\in C}G_z(x,y)\nu(y)$ always exists as an element in ${}^*\mathbf{R}$ (it is the sum of an internal function over a hyperfinite set), it may be infinitely large since $G_z(\cdot,\cdot)$ has a singularity on the diagonal. This, in fact, happens when $d > 3$, in which case λ must be negative and infinitesimal.

The only problem in showing that the standard part ${}^\circ F$ of F is a perturbation of E, is to prove that the two forms are different. If we have chosen λ such that the function h above is negative, finite and noninfinitesimal, it is not hard to check that the difference

(19) $(\tilde{G}_{z_0} - G_{z_0})f = \hat{G}_{z_0}(\frac{1}{\lambda} + G'_{z_0})^{-1}\hat{G}^*_{z_0} f$

is a lifting of a nonzero function for a suitable choice of f (the most "suitable" choice of f would have been $f = \hat{G}^{*-1}_{z_0} h$ if it weren't for the fact that Proposition II-4 requires $E(f,f)$ and $F(f,f)$ to be finite; as it is, one must work with approximations of $\hat{G}^{*-1}_{z_0} h$ instead). The difficulty now is that since it may happen that $z_0 = \inf\{{}^\circ F(f,f) \mid \|f\| = 1\}$, formula (19) is not sufficient for Proposition II-4; what I need is (19) with z_0 replaced by a smaller $z \in \mathbf{R}$. But

(20) $(\tilde{G}_z - G_z)f = \hat{G}_z((\frac{1}{\lambda} + G'_{z_0}) + (G'_z - G'_{z_0}))^{-1}\hat{G}^*_z f,$

and comparing (19) and (20), it is clear that all one needs to require is that $G'_z - G'_{z_0}$ is finite in some appropriate sense. A calculation similar to (17) shows that what is necessary is that the function

$$x \to \sum_{y\in C} (G_z(x,y) - G_{z_0}(x,y))\nu(y)$$

has finite $L^2(C,\nu)$-norm. The last condition is obviously satisfied when $d \leq 3$, because then $\sum_{y\in C} G_z(x,y)\nu(y)$ has finite $L^2(C,\nu)$-norm for all negative $z \in \mathbf{R}$.

When $d > 3$, the quantity $\sum_{y\in C} G_z(x,y)\nu(y)$ is infinite for all real z, and at first glance it may seem improbable that the differences

$$\sum_{y\in C} (G_z(x,y) - G_{z_0}(x,y))\nu(y)$$

can be finite. Observe, however, that by the resolvent equation

$$G_z(x,y) - G_{z_0}(x,y) = (z - z_0)\sum_{u\in\Lambda} G_z(x,u)G_{z_0}(u,y)\delta^d,$$

where the sum over Λ smooths the singularity on the diagonal. In fact, this smoothing is so efficient, that the function $x \to \sum_{y\in C} (G_z(x,y) - G_{z_0}(x,y))\nu(y)$ has finite L^2-norm when $d \leq 5$. We have thus shown that if $d \leq 5$, then ${}^\circ F$ is a perturbation of E supported on B when the function h in (18) is negative, noninfinitesimal and finite. Note that for $d = 4,5$ this implies that λ is negative and infinitesimal. It is not hard check that for $d \leq 3$, the form F in (12) is the standard part of F when λ is a suitable nonstandard representation of $\tilde{\lambda}$.

There is an obvious way of reformulating the idea of this proof in standard terms; instead of working with a hyperfinite random walk on a hyperfinite lattice, I could have used finite walks on finite lattices and taken the limit as the mesh went to zero. That the hyperfinite coupling constant λ is infinitesimal for $d=4$ and 5, would in this approach be reflected by the fact that, as the mesh got finer, I had to let my λ's converge to zero to get convergence of the forms. This, of course, is just an example of the philosophy I was propounding in Part I - that an infinitesimal is nothing but a convenient representation of a sequence converging to zero at a certain rate. The advantage of this representation is that you can do all your constructions and computations on one and the same structure, and that no limit has to be taken (compare the discussion of the nonstandard approach to Brownian motion at the end of Part I).

C. A few speculative remarks on polymer measures and quantum fields.

Formally, the (standard) forms we have constructed are given by operators

$$(21) \quad H = -\frac{1}{2}\Delta + \int_0^1 \lambda(b(s))\,\delta(\cdot - b(s))ds.$$

By the Feynman-Kac formula, the associated semigroup should be given by

$$(22) \quad \langle T_t f, g\rangle = \int f(\tilde{b}(t))g(\tilde{b}(0))e^{-\int_0^t\int_0^1 \lambda(b(s))\delta(\tilde{b}(r)-b(s))dsdr}\,d\tilde{P}$$

where $\tilde{b}$ is a new Brownian motion independent of b. It is possible to give precise meaning to and to prove this formula by using a nonstandard version of the Feynman-Kac formula (see [2]). Westwater [16], [17] has shown that for $d=3$ and λ a positive constant, the existence of the exponential in (22) can be used to construct the Edwards polymer measures formally given by

$$dZ_\lambda = e^{-\int_0^1\int_0^1 \lambda\delta(b(r)-b(s))dsdr}\,dW,$$

where W is Wiener measure. If we could extend Westwater's arguments to the case where $d=4$ and λ is a negative infinitesimal, we would get polymer measures in dimension 4. Through Symanzik's representation this question is intimately related to the nontriviality of ϕ_4^4, and thus the arguments I have presented may be taken to indicate that the "right" choice of the coupling constant in constructive ϕ_4^4-theory is a suitable negative infinitesimal. This, however, is pure speculation; for what is known about nonstandard methods in quantum field theory, see chapter seven of [2].

References

1. S. Albeverio, J.E. Fenstad, R. Höegh-Krohn, W. Karwowski, T. Lindstrøm: In preparation.
2. S. Albeverio, J.E. Fenstad, R. Höegh-Krohn, T. Lindstrøm: Nonstandard methods in stochastic analysis and mathematical physics, Academic Press, to appear.
3. R.M. Anderson: A non-standard representation for Brownian motion and Ito integration, Israel J. Math. 25 (1976), 15-46.
4. N.J. Cutland: Nonstandard measure theory and its applications, Bull. London Math. Soc., 15 (1983), 529-589.
5. M. Davis: Applied nonstandard analysis, Wiley, New York, 1977.
6. J.E. Fenstad: Is nonstandard analysis relevant for the philosophy of mathematics, Synthese (to appear).
7. C.W. Henson, L.C. Moore: Nonstandard analysis and the theory of Banach spaces, in A.E. Hurd (ed): Nonstandard analysis - recent developments, Lecture Notes in Mathematics 983, Springer, 1983, 27-112.
8. H.J. Keisler: Foundations of infinitesimal calculus, Prindle, Weber and Schmidt, 1976.
9. P.A. Loeb: Conversion from nonstandard to standard measure spaces and applications in probability theory, Trans. Amer. Math. Soc., 211 (1975), 113-122.
10. R. Lutz, M. Goze: Nonstandard analysis, Lecture Notes in Mathematics 881, Springer, Berlin - Heidelberg - New York, 1981.
11. W.A.J. Luxemburg: Nonstandard analysis, Lecture notes, Pasadena, 1962.
12. A. Robinson: Non-standard analysis, North-Holland, Amsterdam, 1966.
13. Th. Skolem: Über die Nicht-charakterisierbarkeit der Zahlenreihe mittels endlich oder unendlich vieler Aussagen mit ausschliesslich Zahlenvariablen, Fundamenta Mathematicae, 23 (1934), 150-161.
14. K.D. Stroyan, W.A.J. Luxemburg: Introduction to the theory of infinitesimals, Academic Press, New York, 1976.
15. K.D. Stroyan, J.M. Bayod: Foundations of infinitesimal stochastic analysis, North-Holland (to appear).

16. J. Westwater: On Edwards' model for long polymer chains, Comm. Math. Phys., 72 (1980), 131-174.

17. J. Westwater: On Edwards' model for polymer chains III, Comm. Math. Phys., 84 (1982), 459-470.

HAUSSDORF DIMENSION FOR THE STATISTICAL EQUILIBRIUM OF STOCHASTICS FLOWS

Y. LE JAN (*)

1. Let M be a compact manifold of dimension d, and $(T_n, n \in \mathbb{Z})$ a family of i.i.d random C^2-difféomorphisms of M. For all $m,n \in \mathbb{Z}$, $m < n$, we note S_m^n the composed transformation $T_{n-1} T_{n-2} \cdots T_m$.

Let Q be the associated markovian transition probability kernel :

$$Qf(x) = E(f(T_n x)) \quad \text{for all} \quad n \in \mathbb{Z}, \quad x \in M \quad \text{and} \quad f \in C(M).$$

We shall assume in the following that Q is ergodic and denote its invariant probability measure by $\lambda(dx)$. Then, for any $f \in C(M)$, $n \in \mathbb{Z}$, $p \in \mathbb{N}$

$$M_p(f,n) = \int f(S_{n-p}^n x)\, \lambda(dx)$$

is a bounded martingale in p and it converges a.s. towards a limit $\mu_n(f)$. μ_n is a random probability measure on M. We call it the statistical equilibrium of the flow S at time n. (cf our previous works [4], [5], [6] , [7], where it was defined in the context of continuous flows defined by S.D.E's). The law of μ_n is clearly independent of n and $\mu_n = S_m^n \ \mu_m$ for $m < n$.

Concrete examples on the torus were studied in ([4],[5]) where the statistical équilibrium happened to be a.s diffuse for $d \geq 3$ and a.s. Dirac measure for $d = 1$ or 2.

In [0], Baxendale studies a flow on the sphere where the statistical equilibrium appears to be a.s. a Dirac measure.

In [6], [7] the exemple of isotropic flows on the flat space is studied. (In that case, the invariant measure, which is Lebesgue measure, is infinite, and other problems arise).

The purpose of this work is to give a majoration of the average Haussdorf dimension of μ in terms of the Lyapunov spectrum of the flow. The method is a rather straightforward extension of the argument given by Ledrappier in [2] for classical dynamical systems (cf also references in [2]). See [1] for other applications of methods of classical dynamical systems to the context of stochastic flows.

(*) UNIVERSITE PARIS VI - Laboratoire de Probabilités - 4 place Jussieu - Tour 56 3ème Etage - 75230 PARIS 05

We assume that $E(\sup_{x\in M} (\|DT\|^2(x) + \|D^2T\|(x))$ is finite.

It is satisfied for the flows or stochastic differential equations as well as in the case where T_n is defined by a random choice among a finite set of diffeomorphisms.

Remark : Assume that the flow is symmetric, i.e. that T_n and T_n^{-1} are identically distribued, for each n. Then one can define the statistical equilibrium associated with the inverse flow, $\hat{\mu}_n$. For each $n \in \mathbb{Z}$, A measurable subset of E,

$$\hat{\mu}_n(A) = \lim_{p\uparrow+\infty} \lambda(S_n^{n+p}(A)), \text{ which is a bounded martingale in } p.$$

Clearly, μ_n and $\hat{\mu}_n$ are independent and identically distributed. (They correspond to μ_t^- and μ_t^+ defined in [7] in a more specific framework).

It $\mathcal{M}$ is the set of probability measures an E, μ_n and $\hat{\mu}_{-n}$ are obviously stationnary Markov chains on $\mathcal{M}$.

Moreover the joint process $(\hat{\mu}_n, \mu_n)$ is also Markovian.

(Clearly $\sigma(T_m, m < n \; ; \hat{\mu}_n)$ and $\sigma(T_m, m \geq n \; ; \mu_n)$ are independent given $\sigma(\hat{\mu}_n, \mu_n)$).

2. To give a precise statement, we need a few definitions and comments.

a) For any positive random probability measure $\mu(\omega,dx)$ on M, we set :

$$g(\mu) = \lim_{\delta \downarrow 0} \uparrow \limsup_{\varepsilon \downarrow 0} E(\mathrm{Log}\, N(\varepsilon,\delta,\mu)) / \mathrm{Log}(\tfrac{1}{\varepsilon})$$

where $N(\varepsilon,\delta,\mu)$ is the smallest number of open balls of radius ε we need to cover M up to a set of μ-measure strictly smaller than δ.
Note that we can assume that the balls are centered on a countable dense subset of M without changing $N(\varepsilon,\delta,\mu)$ which is therefore a measurable function of μ.
Note also that $N(\varepsilon,\delta,\mu) \leq c\,\varepsilon^{-d}$ for some constant c. Let $(\Omega,\mathcal{A},P)$ be the probability space on which the T_n are defined. One can assume that there is an ergodic shift θ such that $T_{n+1}(\omega) = T_n(\theta\omega)$.

Set $\mathcal{A}^+ = \sigma(T_n, n \geq 0)$ and $\mathcal{A}^- = \sigma(T_n, n < 0)$. If $\tilde{\theta}(\omega,x) = (\theta\omega, T_0(\omega)x)$, $\tilde{\theta}$ is an ergodic shift on $(\Omega \times M, \mathcal{A}^+ \otimes \mathcal{B}(M), P \otimes \lambda)$, and $\tilde{\theta}^n(\omega,x) = (\theta^n\omega, S_0^n(\omega)x)$

μ_0 is $\mathcal{A}^-$-measurable and independent of all the S_0^n.

We can apply Oseledetz theorem (cf [8], [3]) to the Jacobian matrices $DT_n(\omega,x)$ (one uses a measurable trivialisation of the tangent bundle).

Let $s = (\lambda_1,\lambda_2\ldots\lambda_d)$, $\lambda_1 \geq \lambda_2 \geq\ldots\geq \lambda_d$, be their Lyapunov spectrum. Assume $\sum_{i=1}^{d} \lambda_i < 0$.

Following [2], define $\dim \mathrm{dil}(s) = \sum_{i=1}^{j-1} \frac{\lambda_i - \lambda_j}{\lambda_j}$ with $j = \inf(k, \sum_{i=1}^{k} \lambda_i < 0)$

($\dim \mathrm{dil}(s) = 0$ if $\lambda_1 \leq 0$).

Then we have the following :

Proposition : $g(\mu_0) \leq \dim \mathrm{dil}(s)$.

Remark : It implies that $\lim_{\delta \downarrow 0} \uparrow \liminf_{\varepsilon \downarrow 0} \mathrm{Log}\, N(\varepsilon,\delta,\mu_n) / \mathrm{Log}(1/\varepsilon) \leq \dim \mathrm{dil}(s)$ a.s. for all n.

Indeed this limit is clearly independent of n and the inequality follows from the proposition by the 0-1 law.

We cannot prove the inequality for the lim sup as in [2], and the statement given at the end of [5] seems too strong. Nevertheless, μ_0 is a.s. singular.

3. To prove the proposition, let us first introduce a few general notions, relative to a C^2-difféomorphism of M, denoted by T. We define

$$\delta_1(T) = \sup \frac{d(T_x,T_y)}{d(x,y)}.$$

Let ρ be such that, for all $x \in M$, Exp_x^{-1} is a well defined difféomorphism on the ball $B(x,\rho)$.

Then, let $\delta_2 T$ be the supremum, for $d(x,y) < \rho$ and $d(T_x,T_y) < \rho$ of

$d(T_y, \mathrm{Exp}_{T_x} DT(x)\, \mathrm{Exp}_x^{-1} y) / d^2(x,y)$. $\delta_1(T)$ is controlled by DT and $\delta_2(T)$ by D^2T and DT.

Let μ be a positive measure on M, and $\alpha_1 \geq \alpha_2 \geq \ldots \geq \alpha_\ell$ $\ell \leq d$ be positive numbers.

We say that (T,μ) verifies $H(\alpha_1,\ldots\alpha_\ell,\eta)$ iff, on a set A_η of μ-measure $\geq 1-$ there exist a measurable filtration :

$E_1^x = T_xM \supset E_2^2 \supset \ldots \supset E_\ell^x \supset E_{\ell+1}^x = \{0\}$ such that $\|DT(x)U\| \leq \alpha_k \|U\|$ for all $u \in E_k$, x and k.

Lemma : Choose $r < \frac{\rho}{2} \wedge \frac{\rho}{2\delta_1(T)} \wedge \frac{\alpha_j}{8\delta_2(T)}$ and $j \in \{1,\ldots,\ell\}$. For any $\delta > 0$,

$$N(r\alpha_j, \delta + \eta, T(\mu)) \leq 2 \prod_{i=1}^{j-1} \left[\frac{\alpha_i}{\alpha_j}\right]^+ (4d)^d\, N(r,\delta,\mu)$$

(where $[x]^+ = \inf(n \in \mathbb{N}, n > x)$.

The proof is almost the same as the proof of lemma 3.1 in [2]. We recall it for sake of completeness.

Let $\{B_i, i \in I\}$ be a set of balls of radius r, such that $\mu(\bigcup_{i\in I} B_i) > 1-\delta$, with $|I| \leq 2N(r,\delta,\mu)$. Set $I' = \{i \in I, B_i \cap A_\eta \neq \emptyset\}$. Choose $x_i \in B_i \cap A_\eta$, for all $i \in I'$. Set $B_i' = B(x_i,2r)$; $y_i = Tx_i$. We have $\mu(\bigcup_{i\in I} B_i') > 1-\delta-\eta$.

For any $y \in B_i'$, $d(T_y, \mathrm{Exp}_{y_i} DT(x_i)\mathrm{Exp}_{x_i}^{-1}y) \leq \delta_2(T)\, d(x_i,y)^2 \leq 4\delta_2(T) r^2 \leq \frac{r\alpha_j}{2}$.

Set $C_i = \mathrm{Exp}_{y_i}(DT(x_i)\, \mathrm{Exp}_{x_i}^{-1}(B_i'))$. For all $y \in B_i'$, there exists $z \in C_i$ such that $d(T_y,z) \leq \frac{r\alpha_j}{2}$.

From a covering of C_i by K balls of radius $\frac{r\alpha_j}{2}$, we deduce a covering of $T(B_i')$ by K balls of radius $r\alpha_j$, and therefore

$$N(\frac{r\alpha_j}{2}, \delta + \eta, T(\mu)) \leq K|I'| \leq 2K\, N(r,\delta,\mu).$$

Let us define such a covering.

$\mathrm{Exp}_{x_i}^{-1}(B_i')$ is included in an hypercube

$$D_i = \{v \in T_{x_i}M,\ v = \sum_i^d v_k u_k,\ |v_k| < 2r \text{ with } \|u_k\| = 1 \text{ and } \|DT(x_i)u_k\| \leq \alpha_k\}$$

$\mathrm{Exp}_{y_i}^{-1}(C_i) = DT(x_i)(\mathrm{Exp}_{x_i}^{-1}(B_i'))$ is included in the parallelepiped

$$P_i = DT(x_i)(D_i) = \{w \in T_{y_i}M,\ w = \Sigma v_k\, DT(x)u_k,\ |v_k| < 2r\}.$$

The $DT(x_i)u_k$ are independent vectors of $T_{y_i}M$, with norm less than α_k.

Therefore P_i can be covered by $\prod_{i=1}^{j-1} \left[\frac{\alpha_i}{\alpha_j}\right]^+ (4d)^d$ parallelepipeds with sides of lenght less than $\frac{\alpha_j r}{d}$, which can be imbedded, by Exp_{y_i}, in some ball of radius $\frac{r\alpha_j}{2}$.

4. Some estimates are necessary to carry the proof in the stochastic case.

- From the i.i.d property of the T_n and the chain rule, there exist $K > 0$, such that :

$$E(\delta_1(S_0^n) + \delta_2(S_0^n)) \leq (2K)^n \quad \text{for all } n.$$

Therefore, for all $\eta > 0$, $\exists n_1(\eta)$ such that $\delta_1(S_0^n) + \delta_2(S_0^n) \leq K^n$ for all $n \geq n_1(\lambda)$, outside a set $\mathcal{N}^1(\eta) \subseteq \Omega$, of probability less than η.

- Let $\mu_1 > \mu_2 > \ldots > \mu_\ell$ be the distinct values of the spectrum, $P \otimes \lambda$ and therefore $P(d\omega)\ \mu_0(\omega,dx)$ a.s., there exist a measurable filtration

$$E_1^{x,\omega} = T_x M \supset E_2^{x,\omega} \supset \ldots \supset E_\ell^{x,\omega} \supset E_{\ell+1}^{x} = \{0\}$$

such that $\frac{1}{n} \operatorname{Log}(\| DS_0^n(x)u \| / \|u\|) \to \mu_k$ for all $u \in E_k^{x,\omega} / E_{k+1}^{x,\omega}$.

For any $\chi > 0$, set $A^{\omega,\chi,m} = \{x, \| DS_{0,}^n(x)u\| / \|u\| \leq (1+\chi)^n e^{n\mu_k}$ for all $u \in E_k^{x,\omega}, n \geq m$. Since $E(\lambda(A^{\omega,\chi,n})) = E(\mu_0(A^{\omega,\chi,n}))$ increases to one as $n \uparrow \infty$, for any $\eta > 0$, $\exists n_2(\eta,\chi)$ such that : $E(\mu_0(A^{\omega,\chi,n_2})) \geq 1-\eta^2 \Rightarrow \mu_0(A^{\omega,\chi,n_2}) \geq 1-\eta$ except on a $\mathcal{M}^2_{\chi,\eta} \subseteq \Omega$ of probability less than η.

If $\omega \notin \mathcal{M}^2$ and $n \geq n_2$ $(S_0^n(\omega),\mu_0(\omega))$ verifies $H(\alpha_1 \ldots \alpha_\ell,\eta)$ with $\alpha_k = (1+\chi)^n e^{n\lambda_k}$.

Set : $j = \inf k, \sum_1^k \lambda_i < 0$. Set $r_n = \dfrac{e^{n\lambda_j}}{8K^n}$.

For $n \geq n^1(\eta) \vee n^2(\eta,\chi)$, and $\omega \notin \mathcal{M}^1(\eta) \cup \mathcal{M}^2(\chi,\eta)$ the assumptions of the lemma are satisfied by $(S_0^n(\omega), \mu_0(\omega))$. (We can take $\rho = 1$ without loss of generality).

Therefore, for any $\delta > 0$, $n \geq n_1 \vee n_2$ and $\omega \notin \mathcal{M}^1 \cup \mathcal{M}^2$

$$N(r_n e^{n\lambda_j}(1+\chi)^n, \delta+\eta,\mu_n) \leq 2(4d)^d \prod_{i=1}^{j-1} [e^{n(\lambda_i-\lambda_j)}]^+ N(r_n,\delta,\mu_0)$$

taking the logarithm and the expectation, we get inequality,

$$E(\operatorname{Log}(N(r_n e^{n\lambda_i}(1+\chi)^n,\delta+\eta,\mu_0) \leq n \sum_{i=1}^{j-1} (\lambda_i-\lambda_j)$$

$$+ E(\operatorname{Log}(N(r_n,\delta,\mu_0)) + c(1+n\eta)$$

for same constant c independent of the parameters.

Dividing by n and letting $n \uparrow \infty$, $\delta + r \downarrow 0$, and $\chi \downarrow 0$ we get the inequatility

$$(-2\lambda_j + \operatorname{Log} K)\ g(\mu_0) \leq \sum_{i=1}^{j-1} (\lambda_i-\lambda_j) + (-\lambda_j + \operatorname{Log} K)\ g(\mu_0)$$

which yields the proposition.

REFERENCES :

[0] P. BAXENDALE : Asymptotic behaviour of stochastic flows of difféomorphisms. Two case studies. Preprint (1984).

[1] A. CARVERHILL : Flows of stochastic dynamical systems. Ergodic theory. To appear in "Stochastics".

[2] F. LEDRAPPIER : Some relations between dimension and Lyapunov exponents. Comm. Math. Phys. 81, 229-238 (1981).

[3] F. LEDRAPPIER : Ecole d'Eté de Probabilités de St-Flour (1982). Springer Lecture Notes n° 1097.

[4] Y. LE JAN : Equilibrium state for a turbulent flow of diffusion. Proceedings "Stochastic processes and infinite dimensional analysis" Bielefeld 1983. Pitman Lecture Notes. (To appear in 1985).

[5] Y. LE JAN : Equilibre et exposants de Lyapunov de certains flots browniens. CR Acad. Sci. Paris t. 298, Série I, 361-364 (1984).

[6] Y. LE JAN : On Isotropic Brownian Motions. Preprint (1984).

[7] Y. LE JAN : Exposants de Lyapunov pour les mouvements browniens isotropes. CR. Acad. Sci. Paris t. 299, Série I, 947-949 (1984).

[8] V.I. OSELEDETZ : Multiplicative ergodic theorem. Trans. Moscow Math. Soc. 19, 197-221 (1968).

STOPPING PROBLEMS OF SYMMETRIC MARKOV PROCESSES AND NON-LINEAR VARIATIONAL INEQUALITES

Hideo NAGAI

1 - INTRODUCTION

Let us consider the following variational inequality:

$$\begin{cases} \mathcal{E}_\alpha(w,v-w) + (H(.,w(.)),v-w) \geqq \langle \nu,\tilde{v}-\tilde{w}\rangle \\ \qquad\qquad\qquad\qquad\qquad\qquad\qquad\qquad \forall v \geqq g \text{ a.e.} \\ w \geqq g \quad \text{a.e.} \end{cases} \tag{1.1}$$

where $(F,\mathcal{E})$ is a regular Dirichlet space, g is a given function belonging to F, ν is a difference of positive Radon measures of finite energy integrals, $\alpha > 0$ and

$$H(x,u) = -\sup_z \{f(x,z) - c(x,z)u\}$$

$$f(.,z),\ C(.,z) \in L^2 .$$

Inequality (1.1) corresponds to a stopping problem for the symmetric Markov process associated with $(F,\mathcal{E})$, where killing rates and cost functionals represented by f and c are controlled besides stopping times. We treat the characterization problem in § 3 after considering the existence and uniqueness of the solution of the inequality (1.1) in § 2. § 3 is a generalization of [6] and similar problems have been considered in [3], [5] and [9]. Inequality (1.1), when specialized to the diffusion case with regular coefficients, is reduced to a special form of Hamilton-Jacobi-Bellman inequation (cf. [3], [5], [9]). We however treat general symmetric Markov processes including jump type processes and large classes of obstacles. By employing potential theory of Dirichlet space and Markov processes developed in [4], we establish the relationship between the variational inequality (1.1) and a stopping problem without regularity arguments on the solution of (1.1).

In § 4 we treat a problem in passage to the limit in the case where obstacles converge monotonely. We note that Theorem 4.1, combined with Theorem 3.3, indicates that a pay-off function of our stopping problem can be characterized as a quasi-continuous modification of the solution of the variational inequality even if the obstacle is far from a regular function.

§ 5 treats another problem in passage to the limit in the case where the obstacles g_n converge to a function g in the sense of capacitary integrals.

The results in § 4 and § 5 are generalizations of [1], [2] and [7].

The full proofs of the results will appear elsewhere ([8]), here we shall limit ourselves to comment on some main ideas or sketches of the proofs.

2 - VARIATIONAL PROBLEMS IN TERMS OF DIRICHLET SPACE

Let $m(dx)$ be a non-negative everywhere dense Radon measure on a locally compact Hausdorff space X with a countable base. We denote by (F,ε) a regular Dirichlet space relative to $L^2(dm)$. Let Γ be a separable metric space and $f(x,z)$ and $c(x,z)$ be functions on $X \times \Gamma$ satisfying:

(i) $f(x,.)$ and $c(x,.)$ are continuous for m-a.e.x

(ii) $c(.,z)$ is measurable for each z and there exists M such that $0 \leqq c(x,z) \leqq M$ for each z and m-a.e.x

(iii) $f(.,z) \in L^2(m)$ and there exists $f_0(x) \in L^2(m)$ such that $|f(x,z)| \leqq f_0(x)$ m-a.e. for each z.

We then define a function $H(x,u)$ on $X \times \mathbb{R}^1$ by

$$H(x,u) = -\sup_{z\in\Gamma} \{f(x,z) - c(x,z)u\} \quad . \tag{2.1}$$

We note that $H(x,.)$ is a non-decreasing function and that $H(.,u(.)) \in L^2(m)$ if $u(.) \in L^2(m)$. Therefore, if we define an operator H by

$$H u = H(.,u(.)), \quad u \in L^2(m) \ ,$$

then H is a map from $L^2(m)$ to $L^2(m)$ and has a monotone property:

$$(Hu_1 - Hu_2, u_1 - u_2) \geqq 0, \ u_1,u_2 \in L^2(m) \ .$$

For given positive Radon measures ν_1 and ν_2 of finite energy integrals, we first consider the following equation:

$$\varepsilon_\alpha(u,v) + (Hu,v) = \langle\nu,\tilde{v}\rangle, \quad \forall v \in F \tag{2.2}$$

where $\nu = \nu_1 - \nu_2$ and $\tilde{v}$ denotes a quasi-continuous modification of $v \in F$.

Next, we consider for given function $g \in F$ the following variational inequality

$$\begin{cases} \varepsilon_\alpha(w,v-w) + (Hw,v-w) \geqq \langle\nu,\tilde{v}-\tilde{w}\rangle \\ \qquad\qquad\qquad\qquad\qquad\qquad \forall v \in K_g \\ w \in K_g \end{cases} \tag{2.3}$$

where $K_g = \{w \in F\ ;\ \ v \geqq g \text{ m-a.e.}\}$.

Then we have the following propositions.

Proposition 2.1: For any positive Radom measure ν_1 and ν_2 of finite energy integrals, (2.2) has a unique solution.

Proposition 2.2: For each $g \in F$, (2.3) has a unique solution.

The proof of Proposition 2.1 consists in showing that Banach contraction principle applies to the operator $Tw = U_{\alpha+M}\nu - G_{\alpha+M}(Hw - Mw)$.

For the proof of Proposition 2.2 we employ an approximation method. We consider inductively the following sequence of variational inequalities:

$$\begin{cases} \varepsilon_{\alpha+M}(u_0,v-u_0) \geqq \langle\nu,\tilde{v}-\tilde{u}_0\rangle\ , \quad \forall v \in K_g\ ,\ u_0 \in K_g \\ \varepsilon_{\alpha+M}(u_n,v-u_n) \geqq \langle\nu,\tilde{v}-\tilde{u}_n\rangle - (Hu_{n-1} - Mu_{n-1}\ ,\ v-u_n) \\ n = 1,2,\ldots\ . \end{cases}$$

We can show that the sequence $\{u_n\}$ of the solutions of the above variational inequalities converges to some $u \in F$, which is the solution of (2.3). Uniqueness follows from monotonicity of the operator H .

3 - STOCHASTIC CONTROL OF SYMMETRIC MARKOV PROCESSES

Let $M = (\Omega, \mathcal{B}, \mathcal{B}_t, P_x, X_t)$ be a symmetric Markov process on X and we assume that (F,ε) is regular. Let $f(x,z)$ and $C(x,z)$ be functions satisfying (i), (ii), (iii) in § 2. We moreover assume that $f(.,z)$ and $C(.,z)$ are quasi-Borel functions. Let ν be as in § 2 a difference of two positive Radon measures ν_1 and ν_2 of finite energy integrals and A_t be a continuous additive functional of M corresponding to ν. We consider the following stochastic control problems:

$$u^*(x) = \sup_{\{Z_t\}\in M_\Gamma} E_x\left[\int_0^\infty e^{-\alpha t - \int_0^t C(X_s,Z_s)ds} f(X_t,Z_t)dt\right. \tag{3.1}$$

$$+ \int_0^\infty e^{-\alpha t - \int_0^t C(X_s,Z_s)ds} dA_t \Big]$$

$$w^*(x) = \sup_{\{Z_t\}\in M_\Gamma} E_x \Big[\int_0^\tau e^{-\alpha t - \int_0^t C(X_s,Z_s)ds} f(X_t,Z_t)dt \tag{3.2}$$

$$+ \int_0^\tau e^{-\alpha t - \int_0^t C(X_s,Z_s)ds} dA_t + e^{-\alpha\tau - \int_0^\tau C(X_s,Z_s)ds} g(X_\tau)\Big] ,$$

where g is a given quasi-continuous function belonging to F and $M_\Gamma = \{\{Z_t\}$; $\{Z_t\}$ is a progressively measurable Γ-valued process$\}$.

Then we have the following Theorems.

Theorem 3.1: $u^*(x)$ defined by (3.1) is a quasi-continuous modification of the solution u of the equation (2.2).

Theorem 3.2: $w^*(x)$ defined by (3.2) is a quasi-continuous modification of the solution w of the variational inequality (2.3).

The crucial point of the proof of Theorem 3.1 is the following: Let u be the solution of (2.2), then

$$e^{-\alpha t - \int_0^t C(X_s,Z_s)ds} \tilde{u}(X_t) + \int_0^t e^{-\alpha s - \int_0^s C(X_\tau,Z_\tau)d\tau} C(X_s,Z_s)\tilde{u}(X_s)ds$$

$$+ \int_0^t e^{-\alpha s - \int_0^s C(X_\tau,Z_\tau)d\tau} d\bar{A}_s$$

is a martingale for each $\{Z_t\} \in M_\Gamma$, where

$$\bar{A}_s = A_s - \int_0^t H(X_s,\tilde{u}(X_s))ds .$$

Regarding the proof of Theorem 3.2, we note that the solution w of (2.3) can be written

$$w = U_\alpha \nu - G_\alpha(Hw) + U_\alpha \mu$$

for some positive Radom measure μ of finite energy integral and that a continuous additive functional A_t^μ of (P_x,X_t) corresponds to the Radon measure μ. Then we can show that

$$e^{-\alpha t - \int_0^t C(X_s,Z_s)ds}\tilde{w}(X_t) + \int_0^t e^{-\alpha s - \int_0^s C(X_\tau,Z_\tau)d\tau} C(X_s,Z_s)\tilde{w}(X_s)ds$$

$$+ \int_0^t e^{-\alpha s - \int_0^s C(X_\tau,Z_\tau)d\tau} d\tilde{A}_s$$

is a martingale, where $\tilde{A}_t = A_t + A_t^\mu - \int_0^t H(X_s,\tilde{w}(X_s))ds$.

Further observation concerning the support of the measure μ leads us to the proof of Theorem 3.2.

Remark 3.1: Let us put $Q = \{g : g$ is a quasi-Borel function such that there exists $u \in F$ satisfying $\tilde{u} \geqq g$ q.e.$\}$.

If for $g \in Q$ we set a closed convex subset $\tilde{K}_g$ of F as

$$\tilde{K}_g = \{v \in F ; \quad \tilde{v} \geqq g \text{ q.e.}\} ,$$

then we can see in the same way as in Proposition 2.2 that the following variational inequality (3.3) has a unique solution.

$$\begin{cases} \varepsilon_\alpha(w,v-w) + (Hw,v-w) \geqq \langle \nu, \tilde{v} - \tilde{w} \rangle \\ \qquad\qquad\qquad\qquad\qquad\qquad \forall v \in \tilde{K}_g \\ w \in \tilde{K}_g \end{cases} \tag{3.3}$$

Replacing (2.3) by (3.3) we obtain the following version of Theorem 3.2.

Theorem 3.3: If g is a quasi upper semi-continuous function and $g \in Q$, then the same statement as Theorem 3.2 holds by replacing (2.3) by (3.3).

In the same way as the proof of Theorem 3.2 the solution w of (3.3) can be written

$$w = U_\alpha \nu - G_\alpha(Hw) + U_\alpha \nu$$

for some positive Radon measure. Concerning the arguments on the support of the measure μ we use the fact that we can take a non-increasing sequence of quasi-continuous functions which converges quasi-everywhere to g .

4 - PASSAGE TO THE LIMIT (I)

We consider a problem in passage to the limit in the above stochastic control problem. For a given sequence g_n of functions belonging to Q we put

$$w_n^*(x) = \sup_{\{Z_t\} \in M_\Gamma} E_x\Big[\int_0^\tau e^{-\alpha t - \int_0^t C(X_s,Z_s)ds} f(X_t,Z_t)dt + \int_0^\tau e^{-\alpha t - \int_0^t C(X_s,Z_s)ds} dA_t + e^{-\alpha\tau - \int_0^\tau C(X_s,Z_s)ds} g_n(X_\tau)\Big]. \tag{4.1}$$

Corresponding variational inequality to (4.1) is the following:

$$\begin{cases} \mathcal{E}_\alpha(w_n,v-w_n) + (Hw_n,v-w_n) \geqq \langle \nu,\tilde{v}-\tilde{w}_n\rangle \quad \forall v \in K_{g_n} \\ w_n \in K_{g_n} \end{cases} . \tag{4.2}$$

By Remark 3.1 we can define an operator Φ from Q to F such that $\Phi(g_n) = w_n$, $g_n \in Q$ where w_n is the solution of (4.2). Then we have the following theorem.

Theorem 4.1: Let $\{g_n\}$ be a non-decreasing sequence converging quasi-everywhere to a function $g \in Q$ and assume that $w_n^* = \widetilde{\Phi(g_n)}$ q.e., then w_n^* converges quasi-everywhere to a function w^* such that

$$w^*(x) = \sup_{\{Z_t\}\in M_\Gamma} E_x\Big[\int_0^\tau e^{-\alpha t - \int_0^t C(X_s,Z_s)ds} f(X_t,Z_t)dt + \int_0^\tau e^{-\alpha t - \int_0^t C(X_s,Z_s)ds} dA_t + e^{-\alpha\tau - \int_0^\tau C(X_s,Z_s)ds} g(X_\tau)\Big] . \tag{4.3}$$

Moreover $w^* = \widetilde{\Phi(g)}$.

The proof of the present theorem is based on the fact that a non-decreasing sequence of α-potentials whose energy integrals are boundedly dominated converges strongly to some α-potential.

5 - PASSAGE TO THE LIMIT (II)

We introduce a function space $L(C)$ by

$$L(C) = \{\phi : \phi \text{ is a quasi-continuous function such that there exists } u \in F \text{ satisfying } \tilde{u} \geqq |\phi| \text{ q.e.}\} .$$

And we define $C(\phi)$ and $D(\phi)$ for $L(C)$ as follows

$$C(\phi) = \inf \{\varepsilon_\alpha(u,u) \; ; \; u \in F, \; \tilde{u} \geqq |\phi| \text{ q.e.}\}$$

$$D(\phi) = \int_0^\infty \text{cap} \, \{|\phi| > t\} \, dt^2 .$$

Here "cap" means capacity defined in terms of Dirichlet form ε_α. We remark that $L(C)$ is a Banach space with norm $\sqrt{C(\phi)}$. Denoting by Φ_0 the restriction of the operator Φ to $L(C)$, we can state continuity theorem concerning this map Φ_0. We set $\|v\| = \sqrt{\varepsilon_\alpha(v,v)}$. We have the following theorem and Corollaries.

Theorem 5.1: For g_1 and g_2 belonging to $L(C)$ we have

$$\|\Phi_0(g_1) - \Phi_0(g_2)\|^2 \leqq \sqrt{C(g_1 - g_2)} \, (K_1 \sqrt{C(g_1)} + K_1 \sqrt{C(g_2)}$$

$$+ K_2 \|U_\alpha \nu\| + K_3 \|f_0\|_2).$$

Here K_1, K_2 and K_3 are positive constants independent of g_1 and g_2.

Corollary 5.1: Let w_n^* be a pay-off function defined by (4.1) for given $g_n \in L(C)$. For each n, if $g_n \to g$ in $L(C)$ as $n \to \infty$, then $\|w_n^* - w^*\| \to 0$ as $n \to \infty$, where w^* is a pay-off function defined by (4.3) corresponding to g.

Corollary 5.2: Let $g_n, g \in L(C)$ and assume that $D(g_n - g) \to 0$ as $n \to \infty$, then we have $\|w_n^* - w^*\| \to 0$ as $n \to \infty$.

Let us put $w_1 = \Phi_0(g_1)$ and $w_2 = \Phi_0(g_2)$, then there exist positive Radon measures μ_1 and μ_2 of finite energy integrals such that

$$w_i + G_\alpha(Hw_i) - U_\alpha \nu = U_\alpha \mu_i , \quad i = 1,2.$$

By using monotone property of H we obtain

$$\varepsilon_\alpha(w_1 - w_2 , \; w_1 - w_2) \leq \int |g_1 - g_2| d\mu_1 + \int |g_1 - g_2| d\mu_2 .$$

We prove that $\sqrt{\varepsilon_\alpha(w_i, w_i)}$ as well as $\sqrt{\varepsilon_\alpha(U_\alpha\mu_i, U_\alpha\mu_i)}$ is dominated by $K_1\sqrt{\varepsilon_\alpha(U_\alpha\nu, U_\alpha\nu)} + K_2\sqrt{C(g_1)} + K_3 \|f_0\|_2$ for each i, then we have Theorem 5.1.

Corollary 5.1 is a direct consequence of Theorem 3.2 and Theorem 5.1. For the proof of Corollary 5.2 we observe that $C(\phi) \leq 4D(\phi)$.

ACKNOWLEDGMENT: The author wishes to express his hearty thanks to Professor Dr. S. Albeverio who invited him to the 1st BiBoS symposium and gave him valuable advice to publish this article.

REFERENCES

[1] D.R. Adams: Capacity and the obstacle problem, Appl. Math. Optim. 8 (1981) 39-57.

[2] H. Attouch, C. Picard: Inéquation variationelle avec obstacles et espaces fonctionnels en théorie du potentiel, Applicable Anal. 12 (1981) 287-306.

[3] A. Bensoussan, A. Friedman: Nonlinear variational inequalities and differential games with stopping times, J. of Funct. Anal. 16 (1974) 305-352.

[4] M. Fukushima: Dirichlet forms and Markov processes, North-Holland/Kodansha (1980).

[5] N.V. Krylov: Control of a solution of a stochastic integral equation, Theor. of Prob. and its Appl. 27 (1972) 114-131.

[6] H. Nagai: On an optimal stopping problem and a variational inequality, J. of Math. Soc. of Japan 30 (1978) 303-312.

[7] H. Nagai: Impulsive control of symmetric Markov processes and quasi-variational inequalities, Osaka J. of Math. 20 (1983) 863-879.

[8] H. Nagai: Stochastic control of symmetric Markov processes and non-linear variational inequalities, to appear.

[9] M. Nisio: On non-linear semi-group for Markov processes associated with optimal stopping, App. Math. and Optimization 4 (1978) 143-169.

Author's Address:

Hideo NAGAI

INRIA
Domaine de Voluceau-Rocquencourt
BP. 105
F-78153 LE CHESNAY Cedex / France

Department of Mathematics
Tokyo Metropolitan University
FUKASAWA, Setagaya, Tokyo
Japan

MEAN EXIT TIMES AND HITTING PROBABILITIES OF BROWNIAN MOTION IN GEODESIC BALLS AND TUBULAR NEIGHBORHOODS

Mark A. Pinsky
Mathematics Department
Northwestern University
Evanston, IL 60201

In this lecture I will first survey some recent results on Brownian motion in small geodesic spheres of a Riemannian manifold. Then I will turn to some corresponding questions in tubular neighborhoods of a submanifold.

1. Diffusion processes on manifolds.

We first review the modern formulation of diffusion processes. Let M be a differentiable manifold of dimension n and let A be a second order differential operator on M. A diffusion process $X = (X_t, P_x)$ is said to be generated by A if for every twice differentiable bounded real-valued function f with bounded derivatives

$$t \to f(X_t) - \int_0^t (Af)(X_s)ds$$

is a P_x martingale. This has the following consequences:

(1.1) A is positive semi-definite

(1.2) $$E_x f(X_t) = f(x) + E_x \int_0^t (Af)(X_s)ds \qquad (t > 0)$$

(1.3) For every bounded stopping time T
$$E_x f(X_T) = f(x) + E_x \int_0^T (Af)(X_s)ds$$

The latter is Dynkin's identity [D] which can be thought of as a form of the fundamental theorem of calculus for strong Markov processes. In this direction, we can further replace the integral term by higher powers of the generator and obtain the "stochastic Taylor formula" [AF,AK] written as follows:

(1.4) $$f(x) - E_x f(X_T) = \sum_{k=1}^{N} \frac{(-1)^k}{k!} E_x\{T^k A^k f(X_T)\} + \frac{(-1)^{N+1}}{N!} E_x \int_0^T s^N A^{N+1} f(X_s)ds$$

The function f is supposed to be infinitely differentiable with bounded derivatives.

Dynkin's identity (1.3) immediately yields the probabilistic representation of two classical boundary value problems:

(1.5) $$Af_1 = -1 \text{ in } B, \quad f_1 = 0 \text{ on } \partial B \Rightarrow f_1(x) = E_x(T)$$

(1.6) $$Af_2 = 0 \text{ in } B, \quad f_2 = \delta_y \text{ on } \partial B \Rightarrow f_2(x) = P_x\{X_T \in dy\}/dy$$

Here B is a bounded open set and T is the exit time: $T = \inf\{t > 0: X_t \notin B\}$.

The stochastic Taylor formula (1.4) gives the following probabilistic representation for the biharmonic equation

(1.7) $$A^2 f_3 = 1 \text{ in } B\ , \quad f_3 = 0 = Af_3 \text{ on } \partial B \Rightarrow f_3(x) = \tfrac{1}{2}E_x(T^2)\ .$$

Higher order equations are solved similarly.

In practice one can rarely find an exact solution of these classical equations. Nevertheless Dynkin's formula yields immediately the following representation of approximate solutions:

(1.8) $$Af_\varepsilon = -1 + O(\varepsilon^k) \text{ in } B_\varepsilon\ , \quad f_\varepsilon = 0 \text{ on } \partial B_\varepsilon \Rightarrow f_\varepsilon(x) = E_x(T_\varepsilon)[1 + O(\varepsilon^k)]$$

Here B_ε is a sequence of open sets with $x \in B_\varepsilon$, $\operatorname{diam} B_\varepsilon \downarrow 0$. The observation (1.8) will be systematically exploited to study Brownian motion in small spheres.

2. Mean Exit Time from Small Spheres.

Let (M^n, g) be an n-dimensional Riemannian manifold and $b \in \chi(M)$, a smooth vector field. The differential operator

$$A = \frac{1}{2}\Delta + b \cdot \nabla$$

generates a diffusion process, where Δ is the Laplace-Beltrami operator of the metric g. We note that any second-order strictly elliptic operator may be written in this form for a suitable choice of (b,g).

Let T_ε be the exit time from a ball of radius ε with center at $m \in M$:

$$T_\varepsilon = \inf\{t > 0: d(X_t, m) = \varepsilon\}$$

Here d is the distance function determined by the Riemannian metric. Our first result gives the joint influence of (g,b) on the mean exit time [MP 3].

Theorem 0. *For each $m \in M$ we have for $\varepsilon \downarrow 0$,*

$$E_m(T_\varepsilon) = c_0\varepsilon^2 + c_1\varepsilon^4(\tau_m - 6(\operatorname{div} b + |b|^2)_m) + O(\varepsilon^6)$$

where c_0, c_1 are dimension constants and τ is the scalar curvature of the Riemannian metric.

This is proved by finding a function f_ε satisfying $Af_\varepsilon = -1 + O(\varepsilon^4)$ and then applying Dynkin's formula (1.3); f_ε may be found, for example, by solving an ordinary differential equation in the radial variable and averaging over the unit sphere.

In order to discuss inverse problems in stochastic Riemannian geometry, one must obtain an additional term in the expansion of Theorem 0. Indeed, knowledge of the scalar curvature only suffices to determine the Riemannian metric in dimension two!

To obtain the additional term we restrict attention to the case $b = 0$, for technical reasons only. The following result was proved in [GP].

Theorem 1. *Let* T_ε *be the exit time of the diffusion generated by* $\frac{1}{2}\Delta$ *on a Riemannian manifold* (M^n,g). *Then there exist* $c_i = c_i(n)$ $0 \leq i \leq 5$ *such that for each* $m \in M$ *we have for* $\varepsilon \downarrow 0$

$$E_m(T_\varepsilon) = c_0\varepsilon^2 + c_1\tau_m\varepsilon^4 + \varepsilon^6 [c_2 |R|^2 + c_3 |\rho|^2 + c_4\tau^2 + c_5 \Delta\tau]_m + O(\varepsilon^8)$$

where $R = (R_{iajb})$ *is the Riemann tensor and* $\rho = (\rho_{ij})$ *is the Ricci tensor computed in any orthonormal basis of the tangent space* M_m. *Furthermore* $c_i > 0$ *for* $i \neq 3$ *and* $c_2 + c_3 = 0$.

The inverse problem is formulated as follows:

Conjecture. *If for each* $m \in M$, $E_m(T_\varepsilon) = c_0\varepsilon^2$, *then* (M^n,g) *is locally isometric to* (R^n,g_0).

We have obtained a positive result in low dimensions, as follows:

Corollary 1. *Let* $2 \leq n < 6$. *If for each* $m \in M$, $E_m(T_\varepsilon) = c_0\varepsilon^2 + O(\varepsilon^8)$ *then* (M^n,g) *is locally isometric to* (R^n,g_0).

The algebraic mechanism behind Corollary 1 is apparent when we write the quadratic part of the curvature term as follows for $n > 2$:

$$c_2 |R|^2 + c_3|\rho|^2 = c_2(|W|^2 + \frac{6-n}{n-2} |\rho|^2)$$

where W is the Weyl conformal curvature tensor. The above quadratic form is positive definite if $n < 6$. In fact for $n = 6$ there is a non-trivial null space which is concretely presented as follows:

Example. *Let* $M^6 = S^3 \times H^3$, *the product of a sphere of constant sectional curvature* $= +1$ *with a hyperbolic space of constant sectional curvature* $= -1$. *Then* $E_m(T_\varepsilon) = c_0\varepsilon^2 + O(\varepsilon^{10})$.

This is the simplest of a family of Riemannian symmetric spaces with the indicated asymptotic form of the mean exit time. The other members of this family are all of the form $M = G \times G^c$ where G is a compact Lie group with the bi-invariant metric and G^c is the non-compact dual obtained by complexification. The smallest dimension for which G is non-flat is $n = 3$, hence the above example.

In order to obtain a positive result in higher dimensions, we consider the second moment $E_m(T_\varepsilon)$ for which we have obtained the following asymptotic formula [MP 2].

Theorem 2. *Under the hypotheses of Theorem 1, there exist* $d_i = d_i(n)$, $0 \leq i \leq 5$ *such that*

$$E_m(T_\varepsilon^2) = d_0\varepsilon^4 + d_1\tau_m\varepsilon^6 + \varepsilon^8 [d_2 |R|^2 + d_3 |\rho|^2 + d_4\tau^2 + d_5\Delta\tau]_m + O(\varepsilon^{10})$$

Furthermore $d_i > 0$ *for* $i \neq 3$ *while* $d_2 + d_3 \neq 0$.

This leads to the following affirmative solution to the inverse problem in all dimensions.

Corollary 2. <u>Let</u> $2 \leq n < \infty$. <u>If for each</u> $m \in M$ <u>we have both</u> $E_m(T_\varepsilon) = c_0\varepsilon^2 + O(\varepsilon^8)$ <u>and</u> $E_m(T_\varepsilon^2) = d_0\varepsilon^4 + O(\varepsilon^{10})$ <u>then</u> (M^n,g) <u>is locally isometric to</u> (R^n,g_0).

The method to obtain Theorems 1 and 2 is of use in other problems and will be explained briefly. The exponential mapping $\exp_m$ from the tangent space M_m to the manifold M induces a mapping on smooth functions as follows:

$$(\phi_\varepsilon f)(\exp_m x) = f(x/\varepsilon)$$

This operator is used to write the laplacian of the ball of radius ε in terms of operators on the unit ball $\overline{B}_1$ in M_m as follows:

$$\phi_\varepsilon^{-1} \Delta \phi_\varepsilon = \varepsilon^{-2}\Delta_{-2} + \Delta_0 + \varepsilon\Delta_1 + \ldots + \varepsilon^N\Delta_N + \ldots$$

Here Δ_j is a second order differential operator with polynomial coefficients so that so that Δ_j increases the degree of a polynomial by j. In normal coordinates $(x_1,\ldots,x_n)$ at $m \in M$ we have

$$\Delta_{-2} = \sum_{1=1}^{n} \frac{\partial^2}{\partial x_i^2}$$

$$\Delta_0 = (1/3) \sum_{i,a,j,b} R_{iajb}x_a x_b \frac{\partial^2}{\partial x_i \partial x_j} - (2/3) \sum_{i,a} \rho_{ia}\, x_a \frac{\partial}{\partial x_j}$$

The expressions for Δ_1,Δ_2 are given in [GP]. To obtain the result of Theorem 1 we let

$$f_\varepsilon = \phi_\varepsilon(\varepsilon^2 F_0 + \varepsilon^4 F_2 + \varepsilon^5 F_3 + \varepsilon^6 F_4)$$

where F_0,F_2,F_3,F_4 are obtained by solving the following boundary value problems:

$$\Delta_{-2}F_0 + 1 = 0 \quad \text{in } \overline{B}_1 \;, \quad F_0 = 0 \text{ on } \partial\overline{B}_1$$

$$\Delta_{-2}F_2 + \Delta_0 F_0 = 0 \quad \text{in } \overline{B}_1 \;, \quad F_2 = 0 \text{ on } \partial\overline{B}_1$$

$$\Delta_{-2}F_3 + \Delta_1 F_0 = 0 \quad \text{in } \overline{B}_1 \;, \quad F_3 = 0 \text{ on } \partial\overline{B}_1$$

$$\Delta_{-2}F_4 + \Delta_0 F_2 + \Delta_2 F_0 = 0 \quad \text{in } \overline{B}_1 \;, \quad F_4 = 0 \text{ on } \partial\overline{B}_1$$

It is immediate that $\Delta f_\varepsilon = -1 + O(\varepsilon^8)$. The explicit calculations carried out in [GP] give the values of $F_0(0)$, $F_2(0)$, $F_4(0)$ and hence the result of Theorem 1.

To obtain the result of Theorem 2, we let

$$g = \phi_\varepsilon(\varepsilon^4 G_0 + \varepsilon^6 G_2 + \varepsilon^7 G_3 + \varepsilon^8 G_4)$$

where G_0,G_2,G_3,G_4 are obtained by solving the following boundary value problems:

$$\Delta_{-2}G_0 + F_0 = 0 \quad \text{in } \bar{B}_1 \ , \quad G_0 = 0 \text{ on } \partial\bar{B}_1$$

$$\Delta_{-2}G_2 + \Delta_0 G_0 + F_2 = 0 \quad \text{on } \bar{B}_1 \ , \quad G_2 = 0 \text{ on } \partial\bar{B}_1$$

$$\Delta_{-2}G_3 + \Delta_1 G_0 + F_3 = 0 \quad \text{in } \bar{B}_1 \ , \quad G_3 = 0 \text{ on } \partial\bar{B}_1$$

$$\Delta_{-2}G_4 + \Delta_0 G_2 + \Delta_2 G_0 + F_4 = 0 \quad \text{in } \bar{B}_1 \ , \quad G_4 = 0 \text{ on } \partial\bar{B}_1$$

From this it follows that $\Delta g_\varepsilon = 1 + O(\varepsilon^6)$ in the ball of radius ε while $g_\varepsilon = 0 = \Delta g_\varepsilon$ on the boundary. Appealing to the stochastic Taylor formula (1.4) gives $g_\varepsilon(m) = \frac{1}{2}E_m(T_\varepsilon^2) + O(\varepsilon^{10})$. An explicit calculation carried out in [MP 2] gives the values of $G_0(0)$, $G_2(0)$, $G_4(0)$ and hence the result of Theorem 2.

3. Hitting distribution of small spheres.

The methods of the preceeding sections can be used to locate the "most probable path" of a Brownian motion on a Riemannian manifold. This general line of investigation has been undertaken by many people, under the name of the "Onsager-Machlup formula." The following definitive result was obtained by Y. Takahashi and S. Watanabe [TW].

Let (M_n, g) be an n-dimensional Riemannian manifold with $b \in \chi(M)$ a smooth vector field. Let $\psi, \phi: [0,\infty) \to M$ be two smooth curves and $X = (X_t, P_x)$ the diffusion generated by $A = 1/2\, \Delta + b \cdot \nabla$. Their result shows that

$$\lim_{\varepsilon \downarrow 0} \frac{P\{d(X_s,\phi_s) < \varepsilon \text{ for } 0 \le s \le t\}}{P\{d(X_s,\psi_s) < \varepsilon \text{ for } 0 \le s \le t\}} = \frac{\exp\left[-\int_0^t \mathcal{L}(\phi_s, \dot{\phi}_s)ds\right]}{\exp\left[-\int_0^t \mathcal{L}(\psi_s, \dot{\psi}_s)ds\right]}$$

where the Lagrangian $\mathcal{L}$ is

$$\mathcal{L}(\phi,\dot{\phi}) = \frac{1}{2}\,|\dot{\phi} - b(\phi)|^2 + \frac{1}{2}\operatorname{div} b - (\tau/12)$$

Therefore to look for the "most probable path" we should solve an extremal probem for this Lagrangian. This is already interesting in the case $b \equiv 0$, where we obtain the dynamical equations for the most probable path as follows:

$$\ddot{\phi} = -(1/12)\, \tau(\phi)$$

$$\phi(0) = m \in M \ , \quad \dot{\phi}(0) = 0$$

Starting from rest at a point, we move in the direction of greatest negative curvature, according to this result. Let us now indicate how this same general conclusion may be obtained from analysis of the Brownian motion in a small geodesic sphere.

For this purpose let $f \in C^\infty(S^{n-1})$, a smooth function on the unit sphere and

$\Phi_\varepsilon f$ its image on the manifold, a function which is constant along geodesic rays emanating from $m \in M$. The harmonic measure operator is defined by

$$H_\varepsilon f(m) = E_m\{(\Phi_\varepsilon f)(X_{T_\varepsilon})\} = \int_{S^{n-1}} f(\theta)\, P_m\{\exp_m^{-1}(X_{T_\varepsilon}) \in d\theta\}$$

We have the following result [MP 4]:

Theorem 3. *For each* $m \in M$ *when* $\varepsilon \downarrow 0$

$$H_\varepsilon f(m) = \int_{S^{n-1}} [1 - \bar{c}\varepsilon^3 \nabla_\theta \tau]\, f(\theta)\, \omega(d\theta) + O(\varepsilon^4)$$

where $\bar{c} = \bar{c}(n)$, ∇_θ *is the directional derivative and* ω *is normalized Lebesgue measure on* S^{n-1}.

This shows, at least in a weak sense, that the density of the hitting probability measure has a maximum at $\theta = -\nabla\tau/|\nabla\tau|$ if $\nabla\tau(m) \neq 0$. This agrees with the Onsager-Machlup approach to the most probable path.

It is also interesting to compare this result with the non-stochastic mean value operator, discussed by many authors, e.g. [GW]. If dS_ε denotes the surface measure on the geodesic sphere of radius ε, it can be shown [MP 4] that

$$\int_{\partial B_\varepsilon} \Phi_\varepsilon f\, dS_\varepsilon = \int_{S^{n-1}} [1 - \bar{\bar{c}}\varepsilon^3 \nabla_\theta \tau] f(\theta)\, \omega(d\theta) + O(\varepsilon^4)$$

where $\bar{c} \neq \bar{\bar{c}}$. This gives a new proof of the fact that the stochastic and non-stochastic values agree only if the scalar curvature is constant [MP 1].

4. Mean exit time from tubular neighborhoods.

A geodesic ball in a Riemannian manifold is a special case of a tubular neighborhood, defined as follows. Let $P \subsetneq M^n$ be a compact submanifold of dimension q in a Riemannian manifold of dimension n. The tubular neighborhood is

$$P_\varepsilon = \{y \in M^n : d(y,p) \leq \varepsilon\}.$$

(In case $q = 0$, P_ε is a point and we obtain the geodesic ball.) For $0 < q < n$ we may consider the diffusion process in M generated by Δ and its first exit time T_ε from P_ε. The following function is of interest:

$$\varepsilon \to \int_P E_p(T_\varepsilon)\,dp$$

where dp is a Radon measure on P_ε. In particular the case of spheres is noteworthy.

Example. *Let* $P = S^{n-1}(r) \subseteq R^n$. *Then*

$$E_p(T_\varepsilon) = \frac{1}{2}\varepsilon^2 + (n-1)(3-n)\,\varepsilon^4/24R^2 + O(\varepsilon^6)$$

If $n = 3$ *the remainder term is identically zero and thus* $E_p(T_\varepsilon) \equiv \frac{1}{2}\varepsilon^2$.

In order to treat general submanifolds, we parametrize a portion of P_ε by Fermi coordinates (t,a,p) where $-\varepsilon < t < \varepsilon$ $a \in S^{n-q-1}$, $p \in P$. Using these we obtain the following decomposition of the Laplacian

$$\Delta = \frac{\partial^2}{\partial t^2} + \left(\frac{n-q-1}{t} + \frac{\theta_t}{\theta}\right)\frac{\partial}{\partial t} + \Delta_p + t^{-2}\Delta_a + \text{terms containing } \frac{\partial^2}{\partial a_i \partial p_j}$$

Here $\theta = \theta(t,a,p)$ is the volume distortion introduced by the exponential map, Δ_p is a smooth extension the Laplacian of the imbedded submanifold and Δ_a is an operator on S^{n-q-1} with coefficients depending on (t,a,p). The following complete result has been obtained for hypersurfaces [GKP].

Theorem 4. Let $q = n - 1$. Then for each $p \in P$ we have when $\varepsilon \downarrow 0$

$$E_p(T_\varepsilon) = \frac{1}{2}\varepsilon^2 + (\varepsilon^4/24)\,[2\sum_1^{n-1} k_i^2 - (\sum_1^{n-1} k_i)^2 + 2\rho_{NN}] + O(\varepsilon^6)$$

where (k_i) are the principal curvatures of the imbedded hypersurface, and ρ_{NN} is the Ricci curvature in the normal direction.

Of particular interest is the quadratic form

$$Q \doteq 2\sum_1^{n-1} k_i^2 - (\sum_1^{n-1} k_i)^2 = 2|II|^2 - (\mathrm{tr}\ II)^2,$$

where II is the second fundamental form of $P^{n-1} \subseteq M^n$. It is easy to see that the lowest eigenvalue of Q is simple for all $n \geq 2$ and is positive if and only if $n = 2$. The corresponding eigenspace is spanned by "umbilic surfaces," i.e. k_i = const. The other eigenvalue has multiplicity $n - 2$ and the eigenspace is spanned by "minimal surfaces," i.e. $\sum_1^{n-1} k_i = 0$. These remarks can be used to give the following characterization of imbedded spheres in R^n.

Corollary. Let $P \subseteq R^n$ be a hypersurface. Suppose that $\forall \varepsilon > 0$

$$\int_P [E_p^P(T_\varepsilon) - E_p^{S^{n-1}(r)}(T_\varepsilon)]\, dp = 0$$

i) If $n = 2$ and $|P| = 2\pi r$, then $P \cong S^1(r)$;
ii) If $n = 3$, then $P \cong S^2(r)$ for some $r > 0$;
iii) If $n \geq 3$ and $\int_P \tau dp = (n-1)(n-2)/2r^2$, then $P \cong S^{n-1}(r)$.

The case of higher codimension is currently under study.

REFERENCES

[AF] H. Airault and H. Follmer, Relative densities of semimartingales, Inventiones Mathematicae, 27 (1974), 299-327.
[AK] K.B. Athreya and T.G. Kurtz, A generalization of Dynkin's identity, Annals of Probability 1 (1973), 570-579.
[D] E.B. Dykin, Markov Processes, 2 vols., Springer Verlag, New York, Berlin, Heidelberg, 1965.
[GKP] A. Gray, L. Karp, M. Pinsky, The mean exit time from a tubular neighborhood in a Riemannian manifold (in preparation).

[GP] A. Gray and M. Pinsky, The mean exit time from a small geodesic ball in a Riemannian manifold, Bulletin des Sciences Mathematiques 107 (1983), 345-370.

[GW] A. Gray and T.J. Willmore, Mean Value Theorems on Riemannian manifolds, Proc. Royal Soc. Edinburgh 92A (1982), 343-364.

[MP1] M. Pinsky, Moyenne stochastique sur une variete Riemannienne, Comptes Rendus Acad. Sciences. Paris 292 (1981), 991-994.

[MP2] M. Pinsky, Brownian motion in a small geodesic ball, to appear in Asterisque, 1984, Proceedings of the Colloque Laurent Schwartz.

[MP3] M. Pinsky, The mean exit time of a diffusion process from a small sphere, Proceedings of the Amer. Math. Soc., 1984, to appear.

[MP4] M. Pinsky, On non-Euclidean harmonic measure, Annales de l'Institut Henri Poincaré 1985, to appear.

[TW] Y. Takahashi and S. Watanabe, The Onsager-Machlup function of diffusion processes, in Stochastic Integrals, Springer Verlag Lecture Notes in Mathematics, vol. 851 1981, 433-463.

RIGOROUS SCALING LAWS FOR DYSON MEASURES

W.R. Schneider
Brown Boveri Research Center
CH-5405 Baden, Switzerland

1. Introduction

In his pioneering work on disordered harmonic chains Dyson [1] introduced a one-parameter family of probability measures μ_s, $s \geq 0$, as solutions of a homogeneous integral equation. If ρ denotes the probability measure describing the mass disorder of the point masses on the chain, this integral equation may be written as

$$\mu_s([0,x)) = \iint d\rho(y) d\mu_s(z) \chi_{s,x}(y,z) \tag{1.1}$$

where $\chi_{s,x}$ is the characteristic (or indicator) function of the set

$$C_{s,x} = \{(y,z) \in R_+^2 \quad ; \quad sy+z/(1+z) < x\} \tag{1.2}$$

(All measures have support in R_+).

Observable quantities are averages over an ensemble of disordered chains. Mathematically, this corresponds to expectations with respect to the product measure whose factors ρ_n, $n \in Z$, are copies of ρ. In some cases these "infinitely many integrals" reduce to finitely many by introducing Dyson measures, the price to be paid being the task to solve (1.1) (in general, for all $s \geq 0$).

In [2] it was shown that (1.1) has exactly one solution depending vaguely continuously on the parameter s. For s=0 the solution is explicitly obtainable, namely $\mu_0=\delta_0$ (δ_a=Dirac measure supported by a). Hence, crudely speaking, as s approaches zero, the entire probability collapses into the origin. Intuitively, one expects that an s-dependent rescaling of the abscissa may lead to a non-trivial limit as s tends to zero, providing thus a more detailed description for the small s behaviour of μ_s and quantities derived thereof. This scaling approach based on intuition, insight and (numerical) evidence has been remarkably successful [3-6] but remains "without mathematical justification" [7]. A rigorous formulation is presented in Section 2. The proof is split into two parts contained in Sections 3 and 4.

2. The Scaling Laws

Let M be the set of finite measures with support in R_+ satisfying

$$\lambda(R_+) \leq 1 \quad , \quad \lambda \in M \quad . \tag{2.1}$$

The subset $M^1 \subset M$ is characterized by equality in (2.1); the subset $M_o^1 \subset M^1$ is characterized by

$$\lambda(\{0\}) = 0, \quad \lambda \in M_o^1 \quad . \tag{2.2}$$

We set

$$(J\lambda)(x) = \lambda([0,x)) \quad , \quad x > 0 \quad , \tag{2.3}$$

for $\lambda \in M$; it determines λ uniquely.

For $\rho \in M_o^1$ and $s \geq 0$ we define R_s by

$$(JR_s\,\lambda)(x) = \iint d\rho(y)d\lambda(z)\chi_{s,x}(y,z) \tag{2.4}$$

with $\chi_{s,x}$ being the characteristic function of the set $C_{s,x}$ defined in (1.2). Obviously, $\lambda \in M^1$ implies $R_s\lambda \in M^1$; hence, R_s will be considered as a map from M^1 into itself. In [2] it was shown that R_s has a unique fixed point, $\mu_s \in M^1$. Actually, $\mu_s \in M_o^1$ for $s > 0$ and $\mu_o = \delta_o$. Furthermore, μ_s depends vaguely continuously on s; as $\mu_s \in M^1$, this implies also weak continuity [8]. Rescaling is now introduced as follows: Let θ be a continuous increasing function defined on $[0,a)$, $a > 0$, satisfying

$$\lim_{t\downarrow 0} \theta(t)/t = 0 \quad . \tag{2.4}$$

Define the rescaled probability measure μ_t^o associated with μ_s by

$$(J\mu_t^o)(x) = (J\mu_{\theta(t)})(tx) \quad , \quad 0 < t < a \tag{2.5}$$

It is the unique fixed point of R_t^o defined by

$$(JR_t^o\,\lambda)(x) = \iint d\rho(y)d\lambda(z)\chi_{t,x}^o(y,z) \tag{2.6}$$

where $\chi^{o}_{t,x}$ is the characteristic function of

$$C^{o}_{t,x} = \{(y,z) \varepsilon R^2_+ \quad ; \quad \frac{\theta(t)}{t} y + \frac{z}{1+tz} < x \} \quad . \tag{2.7}$$

Note that extending the definition to t=0 yields the relation $JR^{o}_{o}\lambda = J\lambda$, i.e. R^{o}_{o} is the identity.

By construction, μ^{o}_{t} depends weakly continuously on t.

Next we specify three classes $N_k \subset M^1_o$, k=1,2,3, of mass disorder and the associated scaling functions θ_k, k=1,2,3:

(1) $\rho \varepsilon N_1$ iff $\rho \varepsilon M^1_o$ and

$$\int x d\rho(x) = c < \infty \quad ; \tag{2.8}$$

$$\theta_1(t) = c^{-1}t^2 \quad . \tag{2.9}$$

(2) $\rho \varepsilon N_2$ iff $\rho \varepsilon M^1_o$ and

$$\int_M^\infty \left| G(x) - cx^{-1} \right| dx < \infty \tag{2.10}$$

for suitable c > 0, M > 0, where

$$G(x) = 1-(J\rho)(x) \ ; \tag{2.11}$$

$$\theta_2(t) = c^{-1}\, t^2 \,|\ln t|^{-1} \quad . \tag{2.12}$$

(3) $\rho \varepsilon N_3$ iff $\rho \varepsilon M^1_o$ and

$$\lim_{x\to\infty} x^{1-\alpha}\, G(x) = c^{1-\alpha} \tag{2.13}$$

for suitable c > 0, 0 < α < 1;

$$\theta_3(t) = c^{-1}\, t^{(2-\alpha)/(1-\alpha)} \quad . \tag{2.14}$$

Theorem (Scaling Laws). The weak limit μ^{o}_{o} of μ^{o}_{t} for $t\downarrow 0$ exists and is given by

$$\mu^{o}_{o} = \delta_1 \ , \quad k=1,2 \quad ; \quad \mu^{o}_{o} = \mu^{(\alpha)} \ , \quad k=3 \quad , \tag{2.15}$$

where $\mu^{(\alpha)}$ is absolutely continuous with density

$$f^{(\alpha)}(x) = \frac{\beta}{\Gamma(\beta)} \gamma x^{-2} H^{20}_{12}\left(\gamma x^{-1} \middle| \begin{matrix} (-1,1) \\ (-\beta,\beta)(0,\beta) \end{matrix}\right) \tag{2.16}$$

$$\beta = (2-\alpha) \quad , \quad \gamma = (\beta^2\Gamma(\alpha))^\beta \quad .$$

Here, H^{mn}_{pq} is the general Fox function (in current notation, see e.g. [9]) treated in extenso in [10]. Note that Lévy (or stable) distributions are Fox functions [11], a fact seemingly unknown hitherto.

3. Proof - Part 1

To simplify notation we drop the superscript ^{o}. Introducing the Laplace transform L on M

$$(L\lambda)(p) \equiv \tilde{\lambda}(p) = \int e^{-px}\, d\lambda(x) \tag{3.1}$$

we obtain from (2.6), (2.7) the equation

$$\tilde{\mu}_t(p) = \tilde{\rho}(p\theta(t)/t) \int d\mu_t(x)\exp(-\frac{px}{1+tx}) \tag{3.2}$$

for the scaled fixed point μ_t. Subtracting $\tilde{\rho}(p\theta(t)/t)\tilde{\mu}_t(p)$ and dividing by t leads to

$$t^{-1}\{1-\tilde{\rho}(p\theta(t)/t)\}\tilde{\mu}_t(p) = \tilde{\rho}(p\theta(t)/t) \int d\mu_t(x)\, \phi_t(x;p) \tag{3.3}$$

with

$$\phi_t(x;p) = t^{-1}\{\exp(-\frac{px}{1+tx}) - \exp(-px)\} \quad . \tag{3.4}$$

As M is vaguely compact there exist sequences (t_n) with $\lim t_n=0$ such that $(\mu_n) \equiv (\mu_{t_n})$ is convergent. These sequences will be called c-sequences. Different c-sequences may have different limits. The limit μ of a c-sequence may be defective, i.e. $\mu \in M$ but $\mu \notin M^1$ ($1-\mu(R_+)$ is called defect of μ).
Let C_o denote the set of bounded continuous functions on R_+ vanishing at infinity.

Lemma 1. Let (λ_n) be a vaguely convergent sequence in M with limit λ. Let (f_n) be a sequence in C_o converging in supremum norm $\|\ \|$ to $f \in C_o$. Then

$$\lim_{n\to\infty} \int f_n\, d\lambda_n = \int f d\lambda \quad . \tag{3.5}$$

Proof. As $f \in C_0$ there exists to $\varepsilon > 0$ a compact $K \subset R_+$ such that $|f(x)| < \varepsilon$ in its complement. As $\|f-f_n\| < \varepsilon$ for $n > N$ we obtain $|f_n(x)| < 2\varepsilon$ for $x \in R_+ \setminus K$ and $n > N$. There exists a continuous function u with compact support in R_+ satisfying $0 \leq u \leq 1$ and u=1 on K. Hence,

$$\left|\int f(1-u)\,d\lambda\right| < \varepsilon \quad , \quad \left|\int f_n(1-u)\,d\lambda_n\right| < 2\varepsilon \qquad (n > N) \quad . \tag{3.6}$$

By setting

$$\ell_n(g) = \int gud\lambda_n \quad , \quad \ell(g) = \int gud\lambda \tag{3.7}$$

we obtain positive (hence, continuous) linear functionals ℓ_n, ℓ, i.e. elements of the dual B* of B = C(H) with $\| \ \|$, H = support of u (compact), by Riesz theorem. For any Banach space B and its dual B*, it is easily verified that $g_n \to g$ in B, $\ell_n \to \ell$ in B* implies $\ell_n(g_n) \to \ell(g)$, i.e. $|\ell(g) - \ell_n(g_n)| < \varepsilon$ for $n > N'$. Combination with (3.6) yields

$$\left|\int f d\lambda - \int f_n d\lambda_n\right| < 4\varepsilon \quad , \quad n > N'' \quad . \tag{3.8}$$

<u>Lemma 2.</u> The functions $\phi_t(\cdot;p)$, defined in (3.4), belong to C_0. For $t\downarrow 0$ they converge in $\| \ \|$ to $\phi_0(\cdot;p) \in C_0$, where

$$\phi_0(x;\rho) = px^2 \exp(-px) \quad . \tag{3.9}$$

Proof. Fix $x > 0$ arbitrary. Differentiate

$$f(s) = \exp(-u(s)) \quad , \quad u(s) = x/(1+sx) \tag{3.10}$$

twice to obtain

$$f''(s) = g(u(s)) \quad , \quad g(u) = \exp(-u)(u^4-2u^3) \quad . \tag{3.11}$$

By an elementary calculation

$$|g(u)| \leq g(3+\sqrt{3}) \equiv 2C \quad , \quad C = 6(12+7\sqrt{3})e^{-3-\sqrt{3}} \quad . \tag{3.12}$$

Hence,

$$|s^{-1}(f(s) - f(0) - f'(0))| \leq Cs \quad . \tag{3.13}$$

A change of variables and $f'(0) = e^{-x}x^2$ lead to

$$\| \phi_t(\cdot;p) - \phi_o(\cdot;p) \| \leq C\, p^{-2}\, t \tag{3.14}$$

which implies Lemma 2.

Lemma 3. Let $\rho \in N_k$, k=1,2,3. The asymptotic behaviour for $p\downarrow 0$ of the Laplace transform $\tilde{\rho}$ is given by

$$1-\tilde{\rho}(p) \sim \begin{cases} cp & k = 1 \\ cp\,|\ell n\; cp\,| & k = 2 \\ (cp)^{1-\alpha}\,\Gamma(\alpha) & k = 3 \end{cases} \tag{3.15}$$

Proof.

1. Use

$$p^{-1}(1-e^{-px}) \leq \lim_{p\downarrow 0} p^{-1}\,(1-e^{-px}) = x \quad , \tag{3.16}$$

integrate with respect to $\rho \in N$, and apply Lebesgue dominated convergence theorem.

2. Note that

$$p^{-1}(1-\tilde{\rho}(p)) = \tilde{G}(p) = \int G(x)e^{-px}\,dx \tag{3.17}$$

with G defined in (2.11).
Set $G_o(x) = \min(1,c/x)$ and define ρ_o via (2.11); obviously, $\rho_o \in N_2$. As the difference $G-G_o$ is in L^1, it is sufficient to prove $\tilde{G}_o(p) \sim c|\ell n\; cp|$ which follows from $\tilde{G}_o(p) = (1-\exp(-cp))/p + c\, E_1(cp)$ and the asymptotic behaviour of the exponential integral E_1 [12].

3. Application of Theorem 1, p.456 [13] yields $\tilde{G}(p) \sim \Gamma(\alpha)c^{1-\alpha}$, hence (3.15).

Corollary.

$$\lim\; t^{-1}\,\{1-\tilde{\rho}(p\;\theta_k(t)/t)\} = \begin{cases} p & k = 1,2 \\ \Gamma(\alpha)p^{1-\alpha} & k = 3 \end{cases} \tag{3.18}$$

Now, take a c-sequence with limit μ. From (3.3), (3.4), Lemmata 1 and 2, and Corollary we obtain

$$\frac{d^2}{dp^2}\tilde{\mu}(p) = \begin{cases} \tilde{\mu}(p) & k = 1,2 \\ \Gamma(\alpha)p^{-\alpha}\,\tilde{\mu}(p) & k = 3 \end{cases} \tag{3.19}$$

leading to

$$\tilde{\mu}(p) = \begin{cases} A\exp(-p) \\ A\,\tilde{\mu}^{(\alpha)}(p) \end{cases} \tag{3.20}$$

with $0 \leq A \leq 1$ (possibly depending on the c-sequence) where $\mu^{(\alpha)}$ is the probability measure with density $f^{(\alpha)}$ given in (2.16) (solving (3.19), k=3, yields an expression containing a modified Bessel function K_β; a Mellin transformation then leads to (2.16), details in [11]). From (3.20) we obtain

$$\mu = \begin{cases} A\,\delta_1 & k = 1,2 \\ A\,\mu^{(\alpha)} & k = 3 \end{cases}, \tag{3.21}$$

in particular,

$$\mu(\{0\}) = 0 \quad . \tag{3.22}$$

<u>Lemma 4</u>. The value of A in (3.21) is A=1 independently of the choice of the c-sequence.

Proof. In Section 4.

<u>Lemma 5</u>. Let X be a sequentially compact Hausdorff space, a > 0 and f_0: $(0,a) \to X$. If all c-sequences (i.e. $t_n \to 0$, $(F_0(t_n))$ convergent) have the same limit, say x_0, then F: $[0,a) \to X$ with $F(t) = F_0(t)$, $t > 0$, and $F(0) = x_0$ is continuous at t=0; in particular, $t_n \to 0$ implies $F(t_n) \to x_0$.

Proof. By contradiction.

As M is Hausdorff and vaguely compact we may apply Lemma 5 to F_0: $(0,a) \to M$ with $F_0(t) = \mu_t$. As here F_0 is continuous the same is true for its extension F. This completes the proof of the theorem.

4. Proof - Part II

One verifies easily [2] that

$$(J\theta\lambda)(x) = \lambda((x^{-1},\infty)) \tag{4.1}$$

defines a map $\theta: M_o^1 \to M_o^1$. As ρ and μ_t belong to M_o^1 we may define $\tau=\theta\rho$ and $\sigma_t = \theta\mu_t$. The latter is the fixed point of T_t where

$$(JT_t\lambda)(x) = \iint d\tau(y)d\lambda(z)\psi_{t,x}(y,z) \tag{4.2}$$

with $\psi_{t,x}$ being the characteristic function of

$$E_{t,x} = \{(y,z) \in R_+^2 \quad ; \quad (\frac{\theta(t)}{ty} + \frac{1}{z+t})^{-1} < x \} \quad . \tag{4.3}$$

Applying the Stieltjes transformation

$$(S\lambda)(q) = \int (x+q)^{-1}\, d\lambda(x) \qquad , \qquad q > 0 \tag{4.4}$$

to the fixed point equation for σ_t yields

$$\int K_t(x,q)d\sigma_t(x) = 0 \tag{4.5}$$

with

$$\begin{aligned} K_t(x,q) = {} & \frac{1}{(x+q)(x+q+t)} - (\frac{x+t}{x+q+t})^2 \ \theta(t)t^{-2} \\ & \cdot \int d\tau(y)\ (y + \frac{\theta(t)}{t}\frac{q(x+t)}{x+q+t})^{-1} \quad . \end{aligned} \tag{4.6}$$

Lemma 6. For $q \downarrow 0$ the asymptotic behavior of $S\tau$ is given by

$$(S\tau)(q) \sim \begin{cases} c & k = 1 \\ c|\ell n\ cq| & k = 2 \\ \frac{\pi(1-\alpha)}{\sin\pi\alpha}\ c^{1-\alpha}\ q^{-\alpha} & k = 3 \end{cases} \tag{4.7}$$

Proof.

1. $\tau=\theta\rho$, $\rho \in N_1$ satisfies $\int d\tau(x)x^{-1} = c < \infty$. Hence, the result follows from Lebesgue dominated convergence theorem.

2. A partial integration yields ($\tau(\{0\}) = 0$)

$$(S\tau)(q) = \int (x+q)^{-2} (J\tau)(x)dx \tag{4.8}$$

For $\tau=\theta\rho$, $\rho \in N_2$ the difference $J\tau - J\tau_0$ with $(J\tau_0)(x) = \min(cx,1)$ is in $L^1(x^{-2})$. Hence, the result follows from

$$(S\tau_0)(q) = c \ \ln \frac{1+cq}{cq} \quad . \tag{4.9}$$

3. An elementary calculation yields

$$(S\tau)(q) = q^{-2} \ \ L(1-L\rho)(q^{-1}) \tag{4.10}$$

From (3.15), k=3, and Theorem 1, p.473 [13] we obtain

$$(S\tau)(q) \sim \Gamma(\alpha)\Gamma(2-\alpha)c^{1-\alpha}q^{-\alpha} \tag{4.11}$$

which is (4.7), k=3.

Let C_f be the set of continuous functions on R_+ with a finite value at infinity (i.e. $f \in C_f$ iff there exists $c \in R$ such that $f-c \in C_o$).
The functions $K_t(\cdot,q)$ defined in (4.6) belong to C_f and converge in supremum norm as $t\downarrow 0$ to $K_o(\cdot,q) \in C_f$ where

$$K_o(x,q) = \begin{cases} (x+q)^{-2} \ (1-x^2) & , \quad k = 1,2 \\ (x+q)^{-2}\{1 - \frac{\pi(1-\alpha)}{\sin\pi\alpha} \ x^{2-\alpha} \ (\frac{x+q}{q})^{\alpha}\} & , \quad k = 3 \end{cases} \tag{4.12}$$

as is seen using Lemma 6.

Lemma 7. Let (λ_n) be a sequence in M^1 converging vaguely to $\lambda \in M^1$. Let (f_n) be a sequence in C_f converging in supremum norm to $f \in C_f$. Then

$$\lim_{n\to\infty} \int f_n \, d\lambda_n = \int f d\lambda \tag{4.13}$$

Proof. By "shifting" f_n and f to C_o and using $\lambda_n(R_+) = \lambda(R_+) = 1$ the problem is reduced to Lemma 1.

'or any c-sequence μ_t converges vaguely to μ given by (3.21). Hence $\sigma_t=\theta\mu_t$ con-
erges vaguely to

$$\sigma = (1-A)\delta_o + A \cdot \begin{cases} \delta_1 & , \quad k = 1,2 \\ \theta\mu^{(\alpha)} & , \quad k = 3 \end{cases} ; \tag{4.14}$$

ue to (3.22) σ is nondefective. Application of Lemma 7 yields

$$\int K_o(x,q)d\sigma(x) = 0 \quad . \tag{4.15}$$

'rom (4.12) we obtain

$$K_o(0,q) = q^{-2} \quad . \tag{4.16}$$

'he equations (4.14) - (4.16) are compatible only for A=1, q.e.d..

eferences

1] Dyson, F.J.: The dynamics of a disordered linear chain. Phys.Rev. 92, 1331-1338 (1953).
2] Schneider, W.R.: Existence and uniqueness for random one-dimensional lattice systems. Commun.Math.Phys. 87, 303-313 (1982).
3] Bernasconi, J., Schneider, W.R., Wyss, W.: Diffusion and hopping conductivity in disordered one-dimensional lattice systems. Z.Phys. B37, 175-184 (1980).
4] Alexander, S., Bernasconi, J., Schneider, W.R., Orbach, R.: Excitation dynamics in random one-dimensional systems, Rev.Mod.Phys. 53, 175-198 (1981).
5] Schneider, W.R., Bernasconi, J.: In: Lecture Notes in Physics, Vol. 153, pp.389-393. Berlin, Heidelberg, New York: Springer 1982.
6] Nieuwenhuizen, Th.M., Ernst, M.H.: Transport and spectral properties of strongly disordered chains, Phys.Rev. B31, 3518-3533 (1985).
7] Kawazu, K., Kesten, H.: On birth and death processes in symmetric random environment, J.Statist.Phys. 37, 561-576 (1984).
8] Bauer, H.: Wahrscheinlichkeitstheorie und Grundzüge der Masstheorie. Berlin, New York: Walter de Gruyter 1974.
9] Raina, R.K., Koul, C.L.: On Weyl fractional calculus. Proc.Amer.Math.Soc. 73, 188-192 (1979).
10] Braaksma, B.L.J.: Asymptotic expansions and analytic continuations for a class of Barnes-integrals. Compos.Math. 15, 239-341 (1963).
11] Schneider, W.R.: Stable distributions are Fox functions. Preprint (1985).
12] Abramowitz, M., Stegun, I.A.: Handbook of Mathematical Functions. New York: Dover 1965.
13] Doetsch, G.: Handbuch der Laplace-Transformation. Basel, Stuttgart: Birkhäuser 1971.

ASYMPTOTIC FREEDOM : A RIGOROUS APPROACH

R. SENEOR

Centre de Physique Théorique de l'Ecole Polytechnique
Plateau de Palaiseau - 91128 Palaiseau - Cedex - France

I - INTRODUCTION

In the last 10 years a lot of progress has been made in the rigorous construction of field theories. If the results obtained up to last year had to deal essentially with "academic models", i.e. scalar models in low dimension, there was in 84 a great progress made when people realized that asymptotically free theories are roughly similar to superrenormalizable field theories the main difference being as we shall see later in the slower rate of convergence of the former ones. I will report on this progress. Up to now it is far from giving us the possibility of controlling* "physical models", I mean by that the non abelien gauge theories (which are asymptotically free), but clearly are did half of the way if we add the fact that T. Balaban at Harvard has developped a program for the control of superrenormalizable gauge theories (i.e. 2 and 3 dimensional gauge theories). The work I will describe is the result of a collaboration with J. Feldman, J. Magnen, V. Rivasseau ([1]). Similar results obtained by a different method have been given by K. Gawedzki and A. Kupianen ([2]).

The idea of the proof is due to the encounter of 2 different methods :

- the phase space cell expansions which originated in J. Glimm and A. Jaffe ([3]) and where extended to the study of superrenormalizable massles field theories ([4]) where they have to be combined with scaled cluster expansions

- the resummation of the most divergent parts of the coupling constant counterterms used by G. 't Hooft ([5]) and V. Rivasseau ([6]) for the construction of planar φ^4 in $d=4$ with the negative sign of the coupling constant.

The result is something one may call a renormalization group expansion in contrast with ([2]) which makes rigorous the Wilson approach.

II - EUCLIDEAN FIELD THEORIES AND ASYMPTOTIC FREEDOM

By complexifying the time $t \to it$ field theories become identical to continuous models of statistical mechanics with unbounded spins. The simplest example

* At the time this talk was written J. Feldman, J. Magnen, V. Rivasseau and the author on one side, K. Gawedzki, A. Kupianen on the other side give a construction of the massive 2-dimensional Gross-Neveu model. This model is reputed to be close to 4-dimensional gauge theory.

being the free massive field in d dimension, $d > 2$, described by a Gaussian measure $d\mu_C$ of covariance

$$C(x-y) = (-\Delta + m_0^2)^{-1}(x,y) \simeq \frac{e^{-m_0|x-y|}}{|x-y|^{d-2}} \tag{II.1}$$

and mean zero.

Roughly speaking

$$d\mu_C(\varphi) \simeq \frac{e^{-\frac{1}{2}\int|\nabla\varphi|^2(x)}}{\int e^{-\frac{1}{2}\int|\nabla\varphi|^2(x)} \prod_x d\varphi(x)} \prod_x d\varphi(x)$$

where the r.h.s. has to be understand as the limit after a discretization of the space $\mathbb{R}^d$.

The objects of interest are the moments of this measure

$$<\prod_{i=1}^{n} \varphi(x_i)>_\mu = \int\varphi(x_1)\ldots(x_n)\, d\mu_C(\varphi)$$

One introduces then the interacting fields by defining a cutoff measure

$$d\nu_\Lambda = \frac{1}{Z_\Lambda} e^{-g\mathcal{L}_\Lambda(\varphi)} d\mu_C(\varphi) \tag{II.2}$$

with $Z_\Lambda = \int e^{-g\,\mathcal{L}_\Lambda(\varphi)} d\mu_C(\varphi)$ and $\mathcal{L}_\Lambda(\varphi)$ generally a local function

$$\mathcal{L}_\Lambda(\varphi) = \int_\Lambda v(\varphi(x))\, d^dx \tag{II.3}$$

Similarly one defines the moments $<.>_{\nu_\Lambda}$ of this measure.

A criterium for the constructibility of the measure i.e. the removal of the space cutoff Λ but also of other cutoff as high momentum cutoff (due to the local singularity of $C(x)$ for $x \sim 0$) which may be necessary for the existence of (II.2), was up to now the fact that the perturbation diagrams :

$$\text{connected part of } \{(\frac{d}{dg})^p < \prod_1^n (x_i)>|_{g=0}\} \tag{II.4}$$

have,but a finite number, a finite limit. This is a caracteristic of superrenormalizable theories.

If there is an infinite number of divergent limits but only for a finite number of values of n then the theory is strictly renormalizable.

We show now some of the diagrams appearing in (II.4) for $V(\varphi) = \varphi^4$

Fig. 1 Some perturbation diagrams

The tadpole appearing in the first diagram is given by

$$C(x-x) = C(0) = \frac{1}{(2\pi)^d} \int \frac{1}{p^2+m_o^2} d^d p$$

which is infinite for $d>2$ unless there is an ultraviolet cutoff. This difficulty is overpassed by replacing the products of fields in the interaction by their Wick ordered products. Introducing an ultraviolet cutoff K then $C(0) = \lim_{K\to\infty} C_K(0)$ with finite $C_K(0)$ and

$$:\varphi_K^2: = \varphi_K^2 - C_K(0) \qquad :\varphi_K^4: = \varphi_K^4 - 6C_K(0)\,\varphi_K^2 + 3(C_K(0))^2$$

for the cutoff fields φ_K. This removal of local singularities (the diagram n=2 p=1 disappears) is sufficient in $d=2$ for φ^4. In $d=3$, still for φ^4, the second diagram of Fig. 1 is divergent, we need therefore to renormalize it, then all other diagrams in (II.4) have finite limits : this theory is superrenormalizable. For $d=4$, the same theory is strictly renormalizable.

A measure of the effective strength of the interaction is given by the 0 momentum value of the connected 4-moment (n=4). More precisely one can introduce the dimensionless quantity :

$$g_{eff} = -m^{d-4} \frac{\int \langle\varphi(x_1)\,\varphi(x_2)\,\varphi(x_3)\,\varphi(0)\rangle_\nu^T\, d^d x_1\, d^d x_2\, d^d x_3}{\left(\int \langle\varphi(x_1)\,\varphi(0)\rangle_\nu^T\, d^d x_1\right)^4} \tag{II.5}$$

where m is the physical mass and the subscript T stands for connected part.

Asymptotic freedom means that the contribution $g(E)$ to g_{eff} coming from momenta of order E tends to 0 as $E\to\infty$. Generically $g(E)$ behaves as $\frac{1}{\mathrm{Ln}\,E}$. This is not the behaviour one expects for φ^4 in $d=4$ unless the coupling constant is negative.

III - A SCALAR MODEL OF ASYMPTOTICALLY FREE FIELD THEORY

As I said before only non abelian gauge theories are asymptotically free

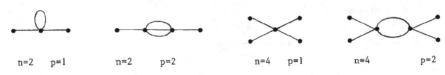

in d=4 on the ultra violet side i.e. for large energy. However φ^4 on the infra red side, i.e. massless φ^4 for small energy is asymptotically free (it is accepted that a theory which is asymptotically free on one side U.V. or I.R. is unstable or singular on the opposite side and vice versa). Let us now describe the model.

For the covariance (or the free measure) we put a U.V. cutoff given for example by $\tilde{C}(p) = \frac{e^{-p^2}}{p^2}$ or $\widetilde{(-\Delta_{\text{lattice}})^{-1}}(p)$ where the tilde stands for Fourier transform.

The interacting measure is then

$$d\nu_\Lambda = \frac{1}{Z_\Lambda} e^{-g\int_\Lambda \varphi^4 + \frac{\mu^2}{2}\int_\Lambda \varphi^2} d\mu_C \tag{III.1}$$

with μ^2 a mass term whose function is to compensate the dynamical mass generated by the φ^4 term. Another choice but equivalent choice is

$$d\nu_\Lambda = \frac{1}{Z_\Lambda} e^{-g\int_\Lambda :\varphi^4: - \frac{m^2}{2}\int_\Lambda :\varphi^2:} d\mu_C \tag{III.2}$$

since because of the U.V. cutoff the tadpoles are not U.V. divergent. For simplicity we will adopt the last point of view. The perturbation expansion of the measure shows divergences. Let $\tilde{G}(p)$ be the Fourier transform of a general 2-point function (moment of order 2). A 2-point function insertion into a propagator line gives

$$\frac{1}{p^2}\tilde{G}(p)\frac{1}{p^2} \simeq \frac{1}{p^4}\tilde{G}(0)$$

for small p, thus unless $\tilde{G}(0) = 0$ and there is no reason for, this generates in $d=4$ an infra red divergence. More generally all diagrams with 2 point subdiagrams are divergent. This shows the necessity of a detailed analysis in momentum space.

We thus introduce a sequence of momentum intervals

$$[1,M^{-1}],\ [M^{-1},M^{-2}],\ \ldots\ ,\ [M^{-(i-1)},M^{-i}],\ \ldots$$

with M a large integer, and a corresponding decomposition of C :

$$\tilde{C}(p) = \sum_i \tilde{C}_i(p)$$

$$\tilde{C}_i(p) = \frac{e^{-p^2 M^{2(i-1)}} - e^{-p^2 M^{2i}}}{p^2}$$

Associated to this decomposition is

$$\varphi = \sum_i \varphi_i$$

where the φ_i's have covariances C_i . The free measure factorizes

$$d\mu_C = \prod_i d\mu_{C_i}$$

but not the interacting one because φ^4 and φ^2 do couple the frequencies.

Finally one introduces apart from the volume cutoff Λ an I.R. cutoff (i.e. the sums over i go from 1 to ρ). The removal of all cutoff is done through a convergent perturbation expansion.

IV - THE EXPANSION

The expansion is the successive combination of 2 types of expansions : a cluster expansion and a frequency coupling expansion.

a) The cluster expansion

It is an expansion which allows to remove the space cutoff Λ . Starting from a given region of space one tests its coupling with other regions using the free propagator. This forms clusters of various sizes. The sum over all cluster decomposition is then convergent if 1) the free propagator is integrable
2) one can extract a small constant per "unit" volume.

To adopt this approach to our case, let us suppose there is only one frequency range in our theory, let say i . The free propagator is C_i and

$$\vec{\nabla}^p C_i(x-y) \simeq M^{-2i-ip} e^{-M^{-i}|x-y|} \tag{IV.1}$$

We choose the "unit" volume to measure the cluster size to be of order M^{4i}, since

$$M^{-4i} \int e^{-M^{-i}|x-y|} d^4y \leqslant Cst \tag{IV.2}$$

This means that for each integration we need a factor λM^{-4i} , λ being a small constant.

The cluster expansion introduces truncated perturbative expansions, i.e. the vertices appearing in estimates are those of the interaction* : $g\varphi^4$, $m^2\varphi^2$, $\beta(\vec{\nabla}\varphi)^2$. The bound (IV.1) says that a field φ_i behaves as M^{-i} and each gradient acting on φ_i gives also M^{-i} , thus

$$g\varphi^4 \sim g\, M^{-4i} \;;\; m^2\varphi^2 \sim m^2 M^{-2i} \;;\; \beta(\nabla\varphi)^2 \sim \beta\, M^{-4i} \tag{IV.3}$$

---

* because of the renormalization discussed in the next section we need to introduce in the measures (III.1) or (III.2) a wave function term $-\beta\int_\Lambda(\vec{\nabla}\varphi)^2$.

Therefore one has convergence of the cluster expansion if g , $m^2 M^{2i}$ and β are small.

However the theory does not reduce to a single frequency.

b) The frequency coupling expansion

One tries to test the coupling between different frequency ranges introducing for example (in the case of the coupling between the first momentum range and the other ones) :

$$F(t_1) = e^{-g\int :(\varphi_1+t_1(\varphi_2+\varphi_3+..))^4: - g\int(1-t_1^4) :(\varphi_2+\varphi_3+...)^4:} \qquad \text{(IV.4)}$$

Then $F(1)$ is the initial term and $F(0)$ corresponds to a factorization of the interaction in two parts one with the first frequency range and the other with the remaining momenta. A similar formula is applied to the mass and the wave function terms. If $E(t_1)$ is the interpolated interaction one then uses

$$E(1) = E(0) + E'(0) + \ldots + \frac{1}{4!} E^{(iv)}(0) + \frac{1}{4!}\int_0^1 (1-t_1)^4 E^{(v)}(t_1)\, dt_1$$

This generates perturbation terms (the cluster expansion too) expressed as moments in φ_1 times product of "external" fields $(\varphi_2+\varphi_3+..)$. For low momentum external fields i.e. for φ_i , $i \gg 1$, the moments of φ_1 appears according to the uncertainty principle as constant pointwise vertices, thus according to IV.a) they have to be small and one has to extract from their external fields M^{-4i}. This is not the case if the number of external legs is less than 4. To get more convergence from the external fields one has to renormalized those terms i.e. to substract from them counterterms (zero momentum values) in order that the difference generates gradients acting on the external fields since a gradient is then equivalent to a factor M^{-i} . Thus one has to add to the exponential in (IV.4) some expression corresponding to the four point counterterms. More precisely one replaces in (IV.4) the second term in the exponential by

$$-g_2 \int(1-t_1^4) :(\varphi_2+\varphi_3+..)^4:$$

with $g_2-g = -\{\text{counterterms for 4 point function built from fields } \varphi_1\}$ (IV.5)

The lowest order contribution to the counterterms corresponds to the bubble in the diagram $n=4$, $p=2$ of Fig. 1. It is given by :

$$g_1^2 \int <:\varphi_1(x)^2: :\varphi_1(y)^2:>_\mu d^4y \simeq g_1^2\, \beta \,\mathrm{Ln}\, M$$

where β is a universal constant. To lowest order (IV.5) gives

$$g_2 \simeq g_1 - \beta \text{ Ln } M \, g_1^2$$

A similar analysis can be done for each momentum range and at range i one introduces

$$g_{i+1} = g_i - \{\text{counterterms made of } \varphi_1, \varphi_2, \ldots, \varphi_i\} \simeq g_i - \beta \text{ Ln } M \, g_i^2 \qquad \text{(IV.6)}$$

with the same β.

Writting (IV.6) as

$$\frac{dg_i}{di} = -\beta \text{ Ln } M \, g_i^2$$

one gets that

$$\frac{1}{g_i} - \frac{1}{g} = \beta \text{ Ln } M \, (i-1)$$

or

$$g_i = \frac{g}{1+\beta(i-1)\text{Ln } M} \qquad \text{(IV.7)}$$

Relation (IV.7) shows that the effective coupling at scale i behaves as $\frac{1}{\text{Ln } M^i}$; this is the meaning of asymptotic freedom for this theory.

The same mechanism has to be used in the mass and wave function case. The leading mass term is the second diagram of Fig. 1, and it can be shown that the effective mass m_i^2 at scale i behaves as $g_i^2 M_i^2$, and the effective wave function coefficient β_i behaves as g.

Finally let us show how is made the renormalization in the case of a 2 point function $G(x,y)$.

One writes

$$\int \varphi(x) \, G(x,y) \, \varphi(y) \, d^4x \, d^4y - \frac{1}{2} \int \varphi^2(x) \, G(x,y) \, d^4x \, d^4y - \frac{1}{2} \int \varphi^2(y) \, G(x,y) \, d^4x \, d^4y$$

$$= \frac{1}{2} \int \varphi(x) - \varphi(y)^2 \, G(x,y) \, d^4x \, d^4y = \frac{1}{2} \int (\nabla\varphi)^2(x) \, (x-y)^2 \, G(x-y) \, d^4x \, d^4y$$

$$+ \text{ higher derivative terms} \qquad \text{(IV.8)}$$

For G with fields of highest index j and for external fields of index i, $i > j$, one loses $(M^j)^2$ because of the $(x-y)^2$ (and the exponential decrease $e^{-M_j|x-y|}$) and one gains $(M^{-i})^2$ because of the 2 gradients. The bad factor is controlled by the fact (simple power counting in M^{-j}) that a unrenormalized 2-point function behaves as M_j^2 .

V - THE RESULTS

The fact that the effective coupling constant tends to zero, as the logarithm of the caracteristic momentum, shows as expected that the theory behaves at small momenta or large distance as a free field theory of covariance the initial covariance time a finite constant depending on g (this can be seen at least from the wave function counterterm β_i which asymptotically tends to be of order g). The n-point functions are those of a free theory plus logarithmic correction terms due to the effective coupling. In fact

$$\sum_{i=1}^{\infty} g_i^p M^{-qi} e^{-M^{-i}|x-y|} \underset{\text{large }|x-y|}{\simeq} \frac{1}{|x-y|^q} \frac{1}{(\mathrm{Ln}|x-y|)^p}$$

for p and $q > 0$.

We can therefore set

<u>Theorem</u> : For g small enough, there exists $m^2(g,\Lambda,\rho)$ such that the following limits exist

$$\lim_{\substack{\Lambda\to\mathbb{R}^4\\ \rho\to\infty}} <\prod_{i=1}^{n} \varphi(x_i)>_\nu = <\prod_{i=1}^{n} \varphi(x_i)>$$

$$\lim_{\substack{\Lambda\nearrow\mathbb{R}^4\\ \rho\to\infty}} m(g,\Lambda,\rho) = m(g)$$

Moreover

$$<\varphi(x_1)\,\varphi(x_2)> \underset{|x_1-x_2|\to\infty}{\simeq} \frac{a(g)}{|x-y|^2}\left(1 + O\left(g,\frac{1}{\mathrm{Ln}|x-y|}\right)\right)$$

with $a(g) \simeq 1+O(g)$

$$<\varphi(x_1)\,\varphi(x_2)\,\varphi(x_3)\,\varphi(x_4)>^T = <\varphi(x_1)\,\varphi(x_2)\,\varphi(x_3)\,\varphi(x_4)> -$$

$$- <\varphi(x_1)\,\varphi(x_2)><\varphi(x_3)\,\varphi(x_4)> - <\varphi(x_1)\,\varphi(x_3)><\varphi(x_2)\,\varphi(x_4)>$$

$$- <\varphi(x_1)\,\varphi(x_4)><\varphi(x_2)\,\varphi(x_3)> \leq \mathrm{Cst} \int \prod_{i=1}^{4} <\varphi(x_i)\,\varphi(z)> \frac{1}{(1+\inf_{(i,j)} \mathrm{Ln}|x_i-x_j|)}\, d^4z$$

Sketch of the proof of the theorem

At each scale of index i, i=0,1,2,... we perform successively

- a cluster expansion to test the coupling between cubes of $\mathcal{D}_i$. One introduces interpolating parameters $s_{\Delta,\Delta'}$, $0 \leq s_{\Delta,\Delta'} \leq 1$, between pairs of cubes, and interpolating covariances

$$C_i(s_{\Delta,\Delta'};\ x,y) = \sum_{(\Delta,\Delta')\in \mathcal{D}_i} s_{\Delta,\Delta'}[\chi_\Delta(y)\chi_{\Delta'}(y) + \chi_\Delta(y)\chi_{\Delta'}(x)]\ C_i(x,y) \quad +$$

$$+ \sum_{\Delta\in \mathcal{D}_i} \chi_\Delta(x)\chi_\Delta(y)\quad C_i(x,y)$$

with $\mathcal{D}_i$ the set of cubes Δ of volume M^{4i}, $\mathcal{D}_i$ being a refinement of $\mathcal{D}_{i-1}$, i=2,3...; χ_Δ is the characteristic function of the cube Δ. The cluster expansion is thus a first order Taylor expansion in each $s_{\Delta,\Delta'}$.

- a frequency coupling expansion which test the coupling between high and low momentum fields i.e. fields of indices less or equal to i for high momentum ones and of indices bigger than i for low momentum. It is realized by the 5th order Taylor expansion of section IV.b).

Each unnormalized Schwinger function S is then a sum of products of connected graphs K(G) of support G :

$$S = \sum_n \sum_{G_1 \ldots G_n} K(G_1)\ldots K(G_n)$$

Two cubes Δ and Δ' of G are connected if a) being in the same $\mathcal{D}_i$ they are linked by a cluster propagator

b) being in different $\mathcal{D}_i$'s they have a vertex in common or they are coupled by the frequency expansion (i.e. linked by a t-variable $\neq 0$). The global connectedness is then the result of the above definition extended by a chain rule.

A general Kirkwood-Salzburg argument shows that the normalized Schwinger functions exist if

$$\sum_{\Delta\in G} |K(G)| < e^{-K|G|} \qquad \text{V.1}$$

for some K large enough.
The sum is over all graphs G containing a fixed cube Δ and $|G|$ is the number of

cubes forming G. This is equivalent to show that one obtains through the expansion a small factor per cube. The sum in V.1 is controlled "horizontally" by the bound IV.2 which allows to sum over distinct cubes and "vertically" (i.e. on the different frequences) by the fact that there is an exponential (vertical) decrease

$$M^{-(\ell_e - 4)(i-j)} \tag{V.2}$$

for each subgraph made of cubes of $\mathcal{D}_1 \cup .. \cup \mathcal{D}_j$ and with ℓ_e external legs in $\mathcal{D}_i$ $(i > j)$. In the case $\ell_e = 2$ or 4, because of the renormalization, we have extra factors $M^{-i}\, M^{j}$ (see IV.6) associated to each gradients, thus finally $\ell_e > 5$. The small factor per cubes has then 2 origins, either the fact that $\ell_e > 5$ is more that what is necessary to sum (V.2) over all $i \geqslant j$, or the fact that it remains from each vertex a $g^{1/4}$. This last point comes from the fact that low momemtum fields, which cannot be gaussianly integrated, are dominated by the exponential of the integration using

$$\left(\frac{g^{1/4}}{|\Delta|} \int_\Delta |\varphi(x)|\, dx\right)^P e^{-g\int_\Delta \varphi^4(x)dx} \leqslant \frac{1}{(|\Delta|)^{P/4}} (P!)^{1/4} \tag{V.3}$$

and that by the form of IV.4 there is at most per vertex 3 low momentum fields.

At each scale i, the approximate value $m^2(i)$ of m^2 is calculated keeping the theory massless, the final value being $\lim_{i\to\infty} m^2(i)$.

The asymptotic freedom has two effects, one is to make the wave function renormalization finite, the second one (from which the first one follows) is to introduce an effective coupling g_i given by IV.7. This last point allows to approximate the Schwinger functions by the lowest order diagrams (giving an effective perturbation theory). In particular the logarithms appearing in the second part of the theorem are due to the insertion of effective vertices (as it is shown by the formula preceding the theorem).

VI. - CONCLUSION

As I said in the introduction, this method can clearly be extended to any asymptotically free field theory provided one overpasses the difficulties linked to the formalism (as for example the degeneracy of the action in gauge theory). It can also be applied to study logarithmic correction to the mean field picture in statistical mechanics models.

REFERENCES

[1] J. Feldman, J. Magnen, V. Rivasseau and R. Sénéor, Proceedings of Les Houches summer school, August-September 1984.

[2] K. Gawedzki, A. Kupianen, Proceedings of Les Houches summer school, August - September 1984 and IHES preprint August 1984.

[3] J. Glim, A. Jaffe, Fort. der Phys. 21, 327 (1973).

[4] J. Magnen, R. Sénéor, Ann. Phys. 152, 130 (1984).

[5] G. 't Hooft, Comm. Math. Phys. 88, 1 (1983).

[6] V. Rivasseau, Comm. Math. Phys. 95, 445 (1984).

The Fermion Stochastic Calculus I

R.F. Streater
Department of Mathematics,
King's College,
Strand, London WC2R 2LS.

Contents

Contents of "The Fermion Stochastic Calculus II"

§1. Notation [1]

The 'one-particle space' will be a real separable Hilbert space $\mathcal{K}$ with scalar product $<\ ,\ >$. The complex Hilbert space spanned by anti-symmetric n-tensors $\{f_1 \wedge \ldots \wedge f_n,\ f_j \in \mathcal{K}\}$ will be written $\Lambda^n\mathcal{K}$. The scalar product between these symbols is determined by the usual tensor scalar product:

$$<f_1 \otimes \ldots \otimes f_n,\ g_1 \otimes \ldots \otimes g_n>_{\otimes\mathcal{K}} = \prod_{j=1}^{n} <f_j, g_j>$$

and the identification

$$f_1 \wedge \ldots \wedge f_n = \frac{1}{\sqrt{n!}} \sum_{\pi \in S_n} P(\pi) f_{\pi(1)} \otimes \ldots \otimes f_{\pi(n)}$$

where π is the permutation

$$\pi = \left\{ \begin{matrix} 1 & 2 & \ldots & n \\ \pi(1) & \pi(2) & & \pi(n) \end{matrix} \right\}$$

and the sum runs over the symmetric group S_n. The anti-symmetric Fock space over $\mathcal{K}$, also called Fermion Fock space, is then $\Lambda\mathcal{K} = \bigoplus_{n=0}^{\infty} \Lambda^n\mathcal{K}$. The creation operator for a particle with wave-function $f \in \mathcal{K}$ acts on $\Lambda\mathcal{K}$ with domain $\bigcup_n \Lambda^n\mathcal{K}$ by linear extension of the map $\Lambda^n\mathcal{K} \to \Lambda^{n+1}\mathcal{K}$ given by

$$a^*(f)(f_1 \wedge \ldots \wedge f_n) = f \wedge f_1 \wedge \ldots \wedge f_n$$

It turns out that $a^*(f)$ is bounded, with norm $\|f\|$. The hermitian Fermi field Ψ is then, for each $f \in \mathcal{K}$

$$\Psi(f) = a^*(f) + a(f).$$

When $\mathcal{K}$ is one-dimensional, spanned by a single vector f of norm 1, then $\Lambda\mathcal{K}$ is two-dimensional, spanned by the normalized vacuum Ω_F and f. We identify $\Lambda\mathcal{K}$ with $\mathbb{C}^2$, and $\Omega_F = \begin{pmatrix}1\\0\end{pmatrix}$, $f = \begin{pmatrix}0\\1\end{pmatrix}$, $a^*(f) = \begin{pmatrix}0 & 0\\1 & 0\end{pmatrix}$, $a(f) = \begin{pmatrix}0 & 1\\0 & 0\end{pmatrix}$, $\Psi(f) = \begin{pmatrix}0 & 1\\1 & 0\end{pmatrix}$. We shall denote $a^*(f)$ by b^* in this case, and $a(f)$ by b.

If $\dim \mathcal{K} = n < \infty$ then $\dim \Lambda\mathcal{K} = 2^n$; but if $\dim \mathcal{K} = \infty$, then $\Lambda\mathcal{K}$ is separable, with $\dim \Lambda\mathcal{K} = \infty$.

The operators $a^*(f)$, $a(f)$ obey the <u>CAR over $\mathcal{K}$</u>

$$[a^*(f), a(g)]_+ = \langle f,g\rangle \cdot \mathbb{1} \qquad f,g \in \mathcal{K}$$

$$[a^*(f), a^*(g)]_+ = [a(f), a(g)]_+ = 0$$

the canonical anti-commutation relations. Here $[A,B]_+ = AB+BA$ where A,B are operators. The c*-algebra generated by $\{a(f),\ f \in \mathcal{K}\}$ is known as the CAR algebra. It is of the type known as uniformly hyperfinite. As represented concretely here, operating on $\Lambda\mathcal{K}$, this algebra is irreducible, so it generates (by weak closure) the W*-algebra $\mathcal{B}(\Lambda\mathcal{K})$, the set of all bounded operators on $\Lambda\mathcal{K}$.

The c*-algebra generated by $\{\Psi(f):\ f \in \mathcal{K}\} = \mathcal{C}(\mathcal{K})$ is the <u>Clifford c*-algebra over $\mathcal{K}$</u>. As here, represented concretely as operators on $\Lambda\mathcal{K}$, $\mathcal{C}(\mathcal{K})$ is reducible. For example if $\dim \mathcal{K} = 1$, then $\mathcal{C}(\mathcal{K})$ is generated by the matrices $\begin{pmatrix}1 & 0\\0 & 1\end{pmatrix}$ and $\begin{pmatrix}0 & 1\\1 & 0\end{pmatrix}$, and is therefore abelian. In general, $\mathcal{C}(\mathcal{K})$ is not abelian. If $\dim \mathcal{K} = \infty$ then the W*-algebra generated by $\mathcal{C}(\mathcal{K})$ (e.g. by weak closure) will be denoted $\mathfrak{A}(\mathcal{K})$. It is a factor of type II_1, known as the hyperfinite factor; this means that the algebra possesses a finite trace i.e. a normalized positive linear functional $A \to \omega(A)$ obeying $\omega(AB) = \omega(BA)$. Indeed, the trace is given by the vacuum expectation functional $A \to \langle\Omega_F, A\Omega_F\rangle = \omega(A)$, $A \in \mathfrak{A}(\mathcal{K})$, where Ω_F is the unit vector (unique up to a scalar) spanning $\Lambda^0\mathcal{K} \cong \mathbb{C}$.

The trace provides a Hilbert space norm on $\mathfrak{A}(\mathcal{K})$, using the scalar product $\langle A,B\rangle = \omega(A^*B)$. We denote $(\omega(A^*A))^{\frac{1}{2}}$ by $\|A\|$. Other norms, analogous to the classical p-norms, can also be defined. This set-up has close parallels with the theory of probability measures, and can be justifiably called quantum probability.

Let $\chi_{[0,t]}$ denote the indicator function on the set $[0,t] \subseteq \mathbb{R}$, $t > 0$. Suppose $\mathcal{K} = L^2[(0,\infty), dt]$, where we interpret t as time. Let

$A_t^* = a^*(\chi_{[0,t]})$ and let $\Psi_t = A_t^* + A_t$. We call $(\Psi_t)_{t\geq 0}$ the Clifford process, and it has been the main objects of our study: it is the Fermion analogue of Brownian motion. We hope to justify this assertion in the next section.

§2. Three motivations

1. The damped oscillator with quantum noise. [2].

We want to describe the phenomenology of a single fermion in a noisy environment. An oscillator of angular velocity ω is described by the operator b (of §1) and has time-evolution $b^*(t) = e^{i\omega t-\gamma t}b$, where γ is the damping due to loss of energy to the environment. This solution obeys the equation

$$\frac{db^*(t)}{dt} = i\omega b^*(t) - \gamma b^*(t)$$

All this takes place on the Hilbert space of the system $C^2 \simeq \Lambda\mathbb{R}$. This damped oscillator violates the CAR at each time $t > 0$, since

$$[b^*(t),b(t)]_+ = e^{-2\gamma t}\mathbb{1} \to 0 \text{ as } t \to \infty.$$

Indeed, the term $-\gamma b^*$ in the equation of motion, due to forces between the system and the environment, should be balanced by a 'noisy' term, representing the input of random forces on the particle. These forces must be fermionic in nature, like the other terms in the equation. This suggests that we look at the Fermion Langevin equation

$$\frac{db^*(t)}{dt} = i\omega b^*(t) - \gamma b^*(t) + \sigma\frac{dA_t^*}{dt}$$

Here, $\frac{dA_t^*}{dt}$ exists only in the sense of distributions, and is usually written $a^*(t)$, the Fermi creation operator at time t. It obeys

$$[a^*(t),a(s)]_+ = \delta(t-s).$$

Note that this shows that $a^*(t)$ does not obey a causal equation of motion in t: noise is fresh and independent at each new time. We hope to adjust the constant σ so that the solution $b^*(t)$ obeys the CAR at all time

$$[b^*(t),b(t)]_+ = 1, \quad t \geq 0.$$

Indeed, the equation is linear and so, although it is 'formal' in that operator-valued distributions $a(t),a^*(t)$ enter it, we can solve it explicitly, with initial condition $b^*(t)\big|_0 = b^*$. The equation is to be understood to refer to an operator $b^*(t)$ or $b(t)$ at each time t, acting on the space $C^2 \otimes \Lambda L^2(0,\infty)$. The initial value b^* means the operator $b^* \otimes \mathbb{1}_\Lambda$ in this space, and the noise $a^*(t)$ is short for the operator-distribution $1_{C^2} \otimes \frac{d}{dt} A^*(t)$. The solution can then be written as

$$b^*(t) = e^{i\omega t-\gamma t} b^* + \sigma\int_0^t e^{(i\omega-\gamma)(t-s)} a^*(s)ds$$

which makes sense as operators on $C^2 \otimes \Lambda L^2(0,\infty)$. Indeed, $\int_0^t e^{-(i\omega-\gamma)s}$ $a^*(s)ds$ is just the smeared field $a^*(h)$, $h(s) = \chi_{[0,t]} e^{-(i\omega-\gamma)s}$; h can be regarded as a complex combination of elements of $\mathcal{K}$ and $a^*(h)$ is taken to be C-linear. If we use the local anti-commutation relations between $a^*(s)$, $a(t)$, we find for the CAR

$$[b^*(t),b(t)]_+ = e^{-2\gamma t}(1 + \frac{\sigma^2}{2\gamma}(e^{2\gamma t} - 1))$$

which equals 1 if $\sigma^2 = 2\gamma$. This is a sort of fluctuation-dissipation theorem. We therefore put $\sigma^2 = 2\gamma$.

As time goes by, the operators $b^*(t)$, $b(t)$ generate an algebra isomorphic to the initial algebra $B(C^2)$ of 2×2 matrices; there is no decay of anti-commutation relations; the system, represented at time t, retains its integrity: literally, since the number operator $b^*(t)b(t)$ always has eigenvalues 0, 1, as befits a fermion. But one can see the exponential decay of the initial state, as follows. We represent the noisy environment by the state Ω_F of $\Lambda L^2(0,\infty)$ and the initial state by Φ; then a particle at time t is given by the state $b^*(t)(\Phi \otimes \Omega_F)$, and the decay from time t to t+τ is represented by the overlap

$$\begin{aligned}&\langle \Phi \otimes \Omega_F, b(t+\tau)b^*(t)\Phi \otimes \Omega_F\rangle \\ &= e^{-i\omega\tau-\gamma\tau-2\gamma t}\langle\Phi,b^*b\Phi\rangle + e^{(-i\omega-\gamma)\tau}(1-e^{-2\gamma t})\end{aligned}$$

As $t \to \infty$, the initial state Φ is forgotten exponentially fast, and the limit settles down to a "stationary process" $B^*_\infty(\tau)$ with two-point function

$$\langle B_\infty(\tau)B^*_\infty(0)\rangle = e^{(-i\omega-\gamma)\tau}$$

analogous to the Ornstein-Uhlenbeck process. We have succeeded in embedding the original contractive (dissipative) time-evolution $e^{(-i\omega-\gamma)t}$ in a much larger system in which the motion $b^*(t) \to b^*(t+\tau)$ is unitary (since any two copies of $B(C^2)$ are unitarily equivalent).

By the way, similarly our solution $b^*(t)$, $b(t)$ provide a representation of a spin ½-particle in a noisy environment by the identification

$$\sigma_1 = i(b^* - b), \quad \sigma_2 = b + b^*, \quad \sigma_3 = 2 - 2b^*b.$$

The description of a heat bath, rather than the noisy environment used here, is more subtle and needs noise with time correlations. Senitzky [3] has obtain a useful class of stationary processes (for bosons), while some non-stationary processes are obtained in [2], for bosons and fermions. The study of quantum noise without time-correlations was advocated by Lax [4].

Our second motivation is to provide a theory of stochastic quantization for fermions. Whereas the advocates of stochastic quantization, following Nelson [5] probably hope to describe Fermionic behaviour somehow by using classical probability theory, there remains a need to understand what to do to set up a 'Euclidean programme' for Fermions. For Bosons, Guerra and Ruggiero [6] have interpreted Euclidean quantum field theory as stochastic mechanics. Parisi and Wu [7] have suggested that a practical way to arrive at the Gibbs state is to introduce a time-dependent stochastic equation and follow it for large times. For linear bosons and fermions, indeed our equations [2] do lead, for large times, to quantum equilibrium states. Our solution to the damped oscillator does suggest the following mechanism for solving a quantum theory: the long time limit of the two-point function of the solution b(t) is

$$\langle b(t+\tau)b^*(t)\rangle \sim e^{-i\omega\tau-\gamma\tau}, \quad t \to \infty.$$

The microscopic equation of motion for b(t) has impressed itself on the long-time behaviour, in that the parameters ω and γ show up, even though for large times the field b(t) lies almost entirely in the noise-space $\Lambda L^2(0,\infty)$, and the part, b, originally in the system-space C^2 is exponentially small.

After taking the large-time limit, to get limit fields $b_\infty(\tau)$, $b^*_\infty(\tau)$ defined from the expectation-values by the Gelfand-Segal construction, we can take the further limit $\gamma \to 0$. The answer is the free oscillator of frequency ω, the original Hamiltonian motion before damping and noise.

This suggests the following <u>conjecture</u>. Let $H(\psi_i)$ be a Hamiltonian written in terms of a family of fermion operators ψ_i. Solve the damped equations with noise

$$\frac{d\psi_i}{dt} = i[H,\psi_i(t)] - \gamma\psi_i(t) + \sigma\frac{d\Psi_i}{dt}, \quad i = 1,2,\ldots \quad (S)$$

where Ψ_i is a copy of the Clifford process for each i. Then we should be able to solve the Hamiltonian equations $\frac{d\psi_j}{dt} = i[H,\psi_j]$ by first solving the stochastic equation (S), taking that t = ∞ limit, and then taking $\gamma \to 0$.

Our third motivation is mathematical. It is known that Brownian motion can be embedded in Boson Fock space [8,9]. All concepts can be transferred to Fermi Fock space, giving Fermion analogues. Most theorems on Brownian motion turn out to be even more true when applied to Fermions.

§3 The Clifford Process [1]

Let $(\Psi_t)_{t\geq 0}$ denote the family of operators $\Psi_t = A_t + A_t^*$, $t \geq 0$. Clearly, Ψ_t is self-adjoint (but not an observable because of its Fermionic nature). From the CAR and $A_t^2 = 0 = A_t^{*2}$, we see that $\Psi_t^2 = t\,1$. Hence the eigenvalues of Ψ_t are $\pm\sqrt{t}$. Whereas Brownian motion, B_t, moves on average a distance $\sqrt{t}$ in time t, the Clifford process moves exactly $\sqrt{t}$.

In quantum probability, the probabilistic aspects are given by expectation values. Let

$$\mathbb{E}(\Psi_{t_1}\cdots\Psi_{t_n}) = \langle\Omega_F, \Psi_{t_1}\cdots\Psi_{t_n}\Omega_F\rangle$$

The process (Ψ_t) has independent increments in the sense that if A is in the W*-algebra generated by $\{\Psi_s, t_1 \leq s \leq t_2\}$, and B in that generated by $\{\Psi_s, t_3 \leq s \leq t_4\}$, then $\mathbb{E}(AB) = \mathbb{E}(A)\,\mathbb{E}(B)$ whenever $[t_1,t_2]$ and $[t_3,t_4]$ do not overlap.

The higher correlation functions [10], also known as the cumulants, the connected Wightman functions or the truncated functions, are defined inductively by $\mathbb{E}(\Psi_t)_T = 0$ and:

$$\mathbb{E}(\Psi_{t_1}\cdots\Psi_{t_n}) = \mathbb{E}(t_1\ldots t_n)_T + \sum_I (\pm 1)\,\mathbb{E}(t_{i_1}\ldots t_{i_j})_T\cdots\mathbb{E}(t_{i_\ell}\ldots t_p)_T$$

The sum is over all partitions of $(1\ldots n)$ into parts $(i_1\ldots i_j)\ldots(i_\ell\ldots p)$ each part in natural order; this definition differs from the one in commutative probability theory by the ± sign, which is the parity of the permutation $\begin{pmatrix}1\ldots n\\ i_1\ldots p\end{pmatrix}$.

For the Clifford process, all truncated functions with $n > 2$ vanish. Thus one might say that (Ψ_t) is "Gaussian". But the actual distribution of Ψ_t is Bernouilli: in the state Ω_F, Ψ_t takes the values $\pm\sqrt{t}$ with equal probability.

Let $\mathfrak{A}_t$ be the von Neumann algebra generated by $\{\Psi_s, 0 \leq s \leq t\}$. The family $(\mathfrak{A}_t)_{t\geq 0}$ is the analogue of the 'filtration' of a stochastic process; it is continuous to the right and left:

$$\mathcal{O}\!\mathcal{l}_t = W^*\{\mathcal{O}\!\mathcal{l}_s,\ s < t\} = \text{von Neumann algebra generated by } \{\mathcal{O}\!\mathcal{l}_s,\ s < t\};$$

$$\mathcal{O}\!\mathcal{l}_t = \bigcap_{s>t} \mathcal{O}\!\mathcal{l}_s\ ; \qquad \mathcal{O}\!\mathcal{l} = W^*\{\mathcal{O}\!\mathcal{l}_t,\ t \geq 0\}; \quad \mathcal{O}\!\mathcal{l}_s \subseteq \mathcal{O}\!\mathcal{l}_t \text{ if } s \leq t.$$

We say that $\mathcal{O}\!\mathcal{l}_t$ is the "noise up to time t". We saw, in the model of a damped harmonic oscillator, that the solution b(t) of the Langevin equation was built out of the noise a(s) for $s \leq t$. In the case of the Clifford process, a(t) and a*(t) enter only in the hermitian combination

$$\Psi_t = a(\chi_{[0,t]}) + a^*(\chi_{[0,t]}).$$

We would expect that a solution X_t, of a stochastic differential equation

$$dX_t = F(X_t,t)d\Psi_t + G(X_t,t)dt,$$

will be built out of Ψ_s for $s \leq t$; so if X_t were bounded, we would expect that $X_t \in \mathcal{O}\!\mathcal{l}_t$, i.e. the process (X_t) is adapted to the filtration $(\mathcal{O}\!\mathcal{l}_t)$. If X_t is unbounded, as is more usual, we would want X_t to be affiliated to $\mathcal{O}\!\mathcal{l}_t$; $X_t \,\eta\, \mathcal{O}\!\mathcal{l}_t$; thus, X_t is closed and spectral resolutions of $|X_t|$ lie in $\mathcal{O}\!\mathcal{l}_t$, and writing $X_t = |X_t|U$, the partial isometry U belongs to $\mathcal{O}\!\mathcal{l}_t$. A quantum stochastic process, then, is an adapted family $(X_t)_{t\geq 0}$: $X_t \,\eta\, \mathcal{O}\!\mathcal{l}_t$, $t \geq 0$. An important concept in the classical theory is the conditional expectation. In the quantum case the analogue has been defined by Umegaki [11]. In our case, the Fock vacuum Ω_F defines a central state on $\mathcal{O}\!\mathcal{l}$, that is $\mathbb{E}(AB) = \mathbb{E}(BA)$ for all A and $B \in \mathcal{O}\!\mathcal{l}$. It follows that $\mathcal{O}\!\mathcal{l}$ is a finite von Neumann algebra; in fact, $\mathcal{O}\!\mathcal{l}$ and all the $\mathcal{O}\!\mathcal{l}_t$ are factors of type II_1. There is a unique map from $\mathcal{O}\!\mathcal{l}_t$ to $\mathcal{O}\!\mathcal{l}_s$ $(s \leq t)$, called the conditional expectation, given $\mathcal{O}\!\mathcal{l}_s$ and written $\mathbb{E}(\ |\mathcal{O}\!\mathcal{l}_s)$, with the properties

1) $\mathbb{E}(A^*A|\mathcal{O}\!\mathcal{l}_s) \geq 0$ for all $A \in \mathcal{O}\!\mathcal{l}_t$, i.e. it is positive
2) $\mathbb{E}(AB|\mathcal{O}\!\mathcal{l}_s) = A\mathbb{E}(B|\mathcal{O}\!\mathcal{l}_s)$, $A \in \mathcal{O}\!\mathcal{l}_s$, $B \in \mathcal{O}\!\mathcal{l}_t$
3) $\mathbb{E}(\mathbb{E}(A|\mathcal{O}\!\mathcal{l}_s)) = \mathbb{E}(A)$; $\mathbb{E}(1|\mathcal{O}\!\mathcal{l}_s) = 1$
4) $\mathbb{E}(\mathbb{E}(A|\mathcal{O}\!\mathcal{l}_s)|\mathcal{O}\!\mathcal{l}_s) = \mathbb{E}(A|\mathcal{O}\!\mathcal{l}_s)$ i.e. it is a projection.

It is interesting to find the explicit form for $\mathbb{E}(\ |\mathcal{O}\!\mathcal{l}_s)$ for the Clifford process: a handy reference for these questions is the 'red book' of Evans and Lewis [12].

The expectation $\mathbb{E}$ defines a scalar product on $\mathcal{O}\!\mathcal{l}$ by the formula $\langle A,B\rangle = \mathbb{E}(A^*B)$. This gives us the notion of orthogonality among the monomials $\psi(u_1)\ldots\psi(u_n)$, $u_j \in \mathcal{O}\!\mathcal{l}$. Let us now apply the Gram-Schmidt orthogonalization procedure to these monomials 1, $\psi(u)$, $\psi(u_1)\psi(u_2)$ in turn. (We do not, however, normalize them). Thus, as

$\mathbb{E}(\psi(u)) = \langle\Omega_F, \psi(u)\Omega_F\rangle = 0$, we see that 1 and $\psi(u)$ are already orthogonal. The quadratic $\psi(u_1)\psi(u_2)$ is orthogonal to $\psi(u)$. The part orthogonal to 1 is $\psi(u_1)\psi(u_2) - \mathbb{E}(\psi(u_1)\psi(u_2))$; this is known as the "Wick-ordered" product of $\psi(u_1)$ and $\psi(u_2)$, and is written $:\psi(u_1)\psi(u_2):$ Similarly, we discover that the polynomials resulting from the orthogonalization coincide with the Wick product $:\psi(u_1)\ldots\psi(u_n):$ for each n.

The Wick products span the algebra generated by the $\{\psi(u),\ u \in \mathcal{K}\}$, so we can define the conditional expectation by linearity starting with its definition on Wick products:

$$\mathbb{E}(:\psi(u_1)\ldots\psi(u_n):|\mathfrak{A}_s) = :\psi(\chi_{[0,s]}u_1)\ldots\psi(\chi_{[0,s]}u_n):$$

This turns out to be extendable to $\mathfrak{A}$, and to obey Umegaki's axioms.

The Clifford process is already Wick-ordered: $\Psi_t = \psi(\chi_{[0,t]})$. Thus if $s < t$ we have

$$\begin{aligned}\mathbb{E}(\Psi_t|\mathfrak{A}_s) &= \mathbb{E}(\psi(\chi_{[0,t]})|\mathfrak{A}_s) = \psi(\chi_{[0,s]}\chi_{[0,t]}) \\ &= \psi(\chi_{[0,s]}) \qquad\qquad = \Psi_s.\end{aligned}$$

This is the statement that (Ψ_t) is a <u>martingale</u>.

Segal and Nelson [13] define "non-commutative L_p spaces" associated with a type II_1 factor, a, as follows. Let $A \,\eta\, \mathfrak{A}$ be "measurable" in a technical sense, and suppose $\mathbb{E}(|A|^p) < \infty$. Then we say $A \in L^p(\mathfrak{A},\mathbb{E})$. If $A \in \mathfrak{A}$ it is always measurable; we write $A \in L^\infty(\mathfrak{A},\mathbb{E})$ in this case. Most classical theorems on L^p-spaces have analogues: the spaces $L^p(\mathfrak{A})$ are decreasing as p increases; they are complete, and L^p, L^q dual, $1 < p < \infty$ if $\frac{1}{p} + \frac{1}{q} = 1$. Moreover, Hölder inequalities can be proved [14]. The main use of this theory here is that operators in L^p can be added and multiplied (strong addition, strong multiplication) in that they have a common core such that their sum on the core is essentially closed, etc. Thus, we do not need to worry about domains. In this notation, Segal's duality map states that there is a natural isomorphism D between $L^2(\mathfrak{A}_t, \mathbb{E})$ and the Fock space $\Lambda(L^2(0,t))$. This isometry is given by $A \to A\Omega_F = DA$, $A \in \mathfrak{A}_t$. The vector Ω_F is then both cyclic and separating for $\mathfrak{A}_t$ (represented on $\Lambda(L^2(0,t))$). Segal's duality map expresses the well-known relation between the Clifford and Grassman algebras (and extends this to infinite dimensions). Since the direct summands in $\Lambda(\mathcal{K}) = \oplus_n \Lambda^n\mathcal{K}$ are orthogonal, we see at once that the orthogonality of the Wick monomials implies that

$$D:\psi(u_1)\ldots\psi(u_n): = u_1 \wedge \ldots \wedge u_n$$

Many questions about the Clifford process can be reduced to the orresponding "one-particle problem" on the space $\mathcal{K}$. Thus, if P is a rojection on $\mathcal{K}$, then $\Lambda^n P$, meaning $u_1 \wedge \dots \wedge u_n \to Pu_1 \wedge \dots \wedge Pu_n$, is a rojection on $\Lambda^n \mathcal{K}$, and it defines by direct sum a projection ΓP on ΛP called the "second quantized operator"). The inverse image, $D^{-1}\Gamma P$, is hen a conditional expectation on $\mathcal{A}$ (the one onto $\mathcal{A}(P\mathcal{K})$). In articular, if $P = \chi_{[0,s]}$ on $L^2(0,\infty)$, then $D^{-1}\Gamma P$ is $\mathbb{E}(|\mathcal{A}_s)$ above, amely, the conditional expectation, given all the noise up to s.

The one dimensional projection P_t onto the vector $\chi_{[0,t]} \in L^2(0,\infty)$ s called the present. The conditional expectation $D^{-1}\Gamma P_t$ will be enoted $\mathbb{E}(|\Psi_t)$; it is the Umegaki conditional expectation onto the on Neumann algebra generated by Ψ_t (t fixed).

With these definitions, (Ψ_t) is a Markov process [15] i.e.

$$\mathbb{E}(\Psi_s|\mathcal{A}_t) = \mathbb{E}(\Psi_s|\Psi_t), \quad \text{if} \quad s \geq t.$$

It is easy to formulate vector-valued Clifford process; we just eed n independent copies $\Psi_t^{(j)}$, $j = 1,2,\dots,n$, each anti-commuting with ll the others. Even more generally, the analogue of a generalized andom field is a quantized Fermi field $\Psi(f,t)$, where f is a test-unction. This would describe the Euclidean version of a neutral field, uch as the Majorana field. Of more interest is the charged field, .e. the Dirac field. The non-relativistic version of this is the omplex Clifford process: let $\mathcal{K}$ now be the real Hilbert space $^2(0,\infty) \oplus L^2(0,\infty)$, and form the Fock space $\Lambda\mathcal{K}$. Let $a(f), a(f)^*$ denote he fermion annihilation and creation operators for vectors of the form $\oplus\ 0$, $f \in L^2(0,\infty)$, and let $b(f)$, $b(f)^*$ denote the 'anti-particle' perators, for vectors $0 \oplus f$. We then define the complex Clifford rocess by

$$\Psi_t^c = a(\chi_{[0,t]}) + b^*(\chi_{[0,t]}).$$

theory similar to (Ψ_t) can be developed for (Ψ_t^c).

4. The Ito-Clifford Integral [1]

In order to make sense of the stochastic equation

$$dX_t = F(X_t,t)d\Psi_t + G(X_t,t)dt$$

e write it in integral form

$$X_t = X(0) + \int_0^t F(X_s,s)d\Psi_s + \int_0^t G(X_s,s)ds$$

and then make sense of the integral $\int \dots d\Psi_s$ following Ito's classical prescription. Here, $X(0)$ is an operator on an initial space not involving the noise. When we solve by Picard's method [16]:

$$X^0(t) = X(0),\ldots,X^{n+1}(t) = X(0) + \int_0^t F(X_s^n,s)\,d\Psi_s + \int_0^t G(X_s^n,s)\,ds$$

we encounter stochastic integrals $\int_0^t f(s)\,d\Psi_s$ where $f(s) = F(X_s^n,s)$ uses the noise up to time t, i.e. f(s) is adapted (it may also depend on $X(0)$). It is thus enough to define the integral for adapted integrands. We first define the integral for elementary adapted functions: a function f(s), zero outside an interval $[t_1,t_2]$ and for $s \in [t_1,t_2]$, f(s) is a constant operator f, affiliated to $\mathfrak{A}_{t_1}$ and in $L^2(\mathfrak{A},\mathbb{E})$. For such f, we define

$$\begin{aligned}\int_0^t f(s)\,d\Psi_s &= f.(\Psi_2-\Psi_1),\ t > t_2 \qquad \Psi_j = \Psi_{t_j}\\ &= f.(\Psi_t-\Psi_1),\ t_1 < t < t_2\\ &= 0 \qquad\qquad ,\ t < t_1\end{aligned}$$

We note that the noise 'points to the future' as in Ito's convention.

We can extend this definition of $\int f(s)\,d\Psi_s$ to any f in the linear span of the elementary functions i.e. to any adapted simple function $f(s) = \sum_j f_j \chi_{[t_j,t_{j+1}]}(s)$. To extend it further, to a more useful class of functions, we need an analytic estimate; there is a Fermion version of the Ito isometry:

The Ito-Clifford Isometry

Let f(s) be an adapted simple function with (operator) values in $L^2(\mathfrak{A},\mathbb{E})$. Then

$$\left\|\int_0^t f(s)\,d\Psi_s\right\|_2^2 = \int_0^t \|f(s)\Omega_F\|^2\,ds.$$

This is in fact an identity in Fock space [17]. A more useful proof uses the fact that (Ψ_t) is a martingale [1]:

$$\left\|\int_0^t f(s)\,d\Psi_s\right\|_2^2 = \sum_{i,j} \mathbb{E}(\Delta\Psi_i f_i^* f_j \Delta\Psi_j)$$

where $\Delta\Psi_i = \Psi_{t_{i+1}} - \Psi_{t_i}$. If $t_i < t_j$ then $\Delta\Psi_j$ is forward in time compared with $\Delta\Psi_i f_i^* f_j$. Hence by properties 3) and 2) of $\mathbb{E}(\cdot|\mathfrak{A}_{t_j})$:

$$\begin{aligned}\mathbb{E}(\Delta\Psi_i f_i^* f_j \Delta\Psi_j) &= \mathbb{E}(\mathbb{E}(\Delta\Psi_i f_i^* f_j \Delta\Psi_j|\mathfrak{A}_j))\\ &= \mathbb{E}(\Delta\Psi_i f_i^* f_j\, \mathbb{E}(\Delta\Psi_j|\mathfrak{A}_j))\\ &= 0 \text{ since } \mathbb{E}(\Delta\Psi_j|\mathfrak{A}_j) = 0,\ (\Psi_t) \text{ being a}\end{aligned}$$

martingale.

imilarly if $t_i > t_j$ the term vanishes, so only the diagonal terms ontribute. Then

$$\left\|\int_0^t f(s)d\Psi_s\right\|_2^2 = \sum_i \mathbb{E}(\Delta\Psi_i f_i^* f_i \Delta\Psi_i)$$

$$= \sum_i \mathbb{E}(f_i^* f_i (\Delta\Psi_i)^2)$$

y the trace property

$$= \sum_i \mathbb{E}(f_i^* f_i)(t_{i+1}-t_i)$$

$$= \int_0^t \|f(s)\Omega_F\|^2 ds$$

ince f(s) is a step function, equal to f_i in the interval $[t_i, t_{i+1}]$. his proves Ito's isometry. This allows us to extend the definition of $\ldots d\Psi_s$ from adapted simple functions in a unique way to the completion $\mathcal{L}$ of this space in the norm $\left(\int_0^t \|f(s)\Omega_F\|^2 ds\right)^{\frac{1}{2}}$, and Ito's isometry holds or the extended integral. We show [1] that $\mathcal{L}$ contains the continuous dapted functions.

We show that a stochastic integral defines a martingale, with zero xpectation [1]. It is just as sensitive to the choice of approximating iemann sums as is found in Brownian motion: if we were to evaluate he integrand f(s) at the later time t_{j+1} instead of the earlier time t_j, in each interval $[t_j, t_{j+1}]$ then we would usually get a different nswer. We use the Ito form for definitions and proofs; if needed, a tratonovich form can be constructed in terms of the Ito form.

The stochastic calculus for the Clifford process differs from the to calculus because Ψ_t and $\Delta\Psi_t$ anti-commute. So if $F(\Psi_t) = \Psi_t^2$ we get

$$F(\Psi_t + \Delta\Psi_t) = (\Psi_t + \Delta\Psi_t)^2$$

$$= \Psi_t^2 + \left[\Psi_t, \Delta\Psi_t\right]_+ + (\Delta\Psi_t)^2$$

$$= \Psi_t^2 + dt$$

which uses $d\Psi_t^2 = dt$, but also shows that Taylor's theorem cannot be used, until after the identity $\Psi_t^2 = t1$ has reduced the degree of the function to a linear one.

This article continues in the Proceedings of the Second Workshop in Quantum Probability, Heidelberg, 1984; Editors L. Accardi, A. Frigerio and W. von Waldenfels; Lecture Notes in Mathematics, vol. 1132, Springer

I thank Professor L. Streit for the hospitality of the BiBos research seminar, University of Bielefeld.

I am indebted to K. Anderson for typing the manuscript.

References

[1] C. Barnett, R.F. Streater and I.F. Wilde. The Ito-Clifford Integral. Jour. Funct. Anal. 48, 172-212 (1982).

[2] R.F. Streater. The damped oscillator with quantum noise. J. Phys. A 15, 1477-1486 (1982).

[3] I.R. Senitzky, Phys. Rev. 119, 670 (1960); ibid A3 421 (1970).

[4] M. Lax. Phys. Rev. 145, 111-129 (1965).

[5] E. Nelson, Dynamical Theories of Brownian Motion. Princeton Lecture Notes, Princeton University Press.

[6] F. Guerra and P. Ruggiero: New interpretation of the Euclidean Markov field in the framework of physical space-time. Phys. Rev. Lett. 31, 1022 (1972).

[7] G. Parisi and Y. Wu. Scientia Sinica 24, 483 (1981).

[8] R.F. Streater, Current Commutation Relations and Continuous Tensor Products. Nuovo Cimento 53, 487 (1968).
R.F. Streater, Current Commutation Relations, Continuous Tensor Products and Infinitely Divisible Group Representations; in Local Quantum Theory (R. Jost, Ed.), Academic Press 1969.

[9] T. Hida; Brownian Motion. Springer Verlag, N.Y., Berlin, Heidelberg 1980.
R.L. Hudson and R.F. Streater. Examples of quantum martingales. Phys. Lett. 85A, 64-66 (1981).
R.L. Hudson and R.F. Streater. Ito's rule is the chain rule with Wick ordering. Phys. Lett. 86A, 277-279 (1981).
R.L. Hudson and R.F. Streater, Non-commutative martingales and stochastic integrals in Fock space; in Lecture Notes in Physics 173, Springer 1981.

[10] D. Mathon and R.F. Streater. Infinitely Divisible Representations of Clifford Algebras. Z. für Wakr. verw. Geb. 20, 308-316 (1971).

[11] H. Umegaki, Conditional Expectation in an Operator Algebra. II. Tohuku Math. J. 8, 86-100 (1956).

[12] D.E. Evans and J.T. Lewis. Commun. Dublin Institute for Advanced Studies A, 24 (1977).

[13] I.E. Segal, A non-commutative extension of abstract integration. Ann. of Math. 57, 401-457 (1953). Ibid, 58, 595-596 (1953).
E. Nelson. Notes on non-commutative integration. Jour. Functl. Anal. 15, 103-116 (1974).

[14] J. Dixmier, Formes linéares sur un anneau d'opérateurs. Bull. Soc. Math. France 81, 9-39 (1953).

5] C. Barnett, R.F. Streater and I.F. Wilde. The Ito-Clifford Integral III: Markov property of solutions to stochastic differential equations. Commun. Math. Phys. 89, 13-17 (1983).

6] C. Barnett, R.F. Streater and I.F. Wilde. The Ito-Clifford Integral II, Stochastic differential equations. J. Lond. Math. Soc.(2), 27, 373-384 (1983).

7] R.F. Streater, Quantum Stochastic Integrals, Acta Physica Austriaca, Suppl.XXVI, 53-74 (1984).

8] C. Barnett, Supermartingales on semi-finite von Neumann algebras. J. Lond. Math. Soc. (2), 24, 175-181 (1981).

9] C. Barnett, R.F. Streater and I.F. Wilde. The Ito-Clifford Integral IV: A Radon-Nikodym Theorem and Bracket Processes. J. Operator Theory, 11, 255-271 (1984).

:0] R.G. Bartle. A general bilinear vector integral. Studia Math. 15, 337-352 (1956).

21] C. Barnett, R.F. Streater and I.F. Wilde. Stochastic integrals in an arbitrary probability gauge space. Math. Proc. Camb. Phil. Soc. 94, 541 (1983).

22] C. Barnett, R.F. Streater and I.F. Wilde. Quasi-free quantum stochastic integrals for the CAR and CCR. J. Functl. Anal. 52, 17-47 (1983).

23] R.F. Streater. Quantum Stochastic Processes. Rome II Conference on Quantum Probability (L. Accardi, A. Frigerio and V. Gorini, Eds.).

24] C. Barnett, R.F. Streater and I.F. Wilde. Quantum Stochastic Integrals under Standing Hypotheses; submitted to J. Lond. Math. Soc.

25] E.J. McShane. Stochastic calculus and stochastic models. Academic Press, N.Y. 1974.

26] D. Stroock and V.S. Varadhan, Multidimensional Diffusion Processes (Springer), 1979.

27] H. Hasegawa and R.F. Streater. Stochastic Schrödinger and Heisenberg equations: a martingale problem in quantum stochastic processes. J. Phys. A. L697-L703 (1983).

28] C. Barnett and T. Lyons. "Stopping non-commutative processes", Imperial College preprint, 1984.